Hadoop + Spark
生态系统操作与实战指南

余辉 著

清华大学出版社
北京

内 容 简 介

本书用于Hadoop+Spark快速上手，全面解析Hadoop和Spark生态系统，通过原理解说和实例操作每一个组件，让读者能够轻松跨入大数据分析与开发的大门。

全书共12章，大致分为3个部分，第1部分（第1~7章）讲解Hadoop的原生态组件，包括Hadoop、ZooKeeper、HBase、Hive环境搭建与安装，以及介绍MapReduce、HDFS、ZooKeeper、HBase、Hive原理和Apache版本环境下实战操作。第2部分（第8~11章）讲解Spark的原生态组件，包括Spark Core、Spark SQL、Spark Streaming、DataFrame，以及介绍Scala、Spark API、Spark SQL、Spark Streaming、DataFrame原理和CDH版本环境下实战操作，其中Flume和Kafka属于Apache顶级开源项目也放在本篇讲解。第3部分（第12章）讲解两个大数据项目，包络网页日志离线项目和实时项目，在CDH版本环境下通过这两个项目将Hadoop和Spark原生态组件进行整合，一步步带领读者学习和实战操作。

本书适合想要快速掌握大数据技术的初学者，也适合作为高等院校和培训机构相关专业师生的教学参考书和实验用书。

本书封面贴有清华大学出版社防伪标签，无标签者不得销售。
版权所有，侵权必究。举报：010-62782989，beiqinquan@tup.tsinghua.edu.cn。

图书在版编目（CIP）数据

Hadoop+Spark生态系统操作与实战指南 / 余辉著. — 北京：清华大学出版社，2017（2021.6重印）
ISBN 978-7-302-47967-3

I. ①H… II. ①余… III. ①数据处理软件—指南 IV. ①TP274

中国版本图书馆CIP数据核字（2017）第207245号

责任编辑：夏毓彦
封面设计：王　翔
责任校对：闫秀华
责任印制：宋　林

出版发行：清华大学出版社
网　　址：http://www.tup.com.cn，http://www.wqbook.com
地　　址：北京清华大学学研大厦A座　　邮　编：100084
社 总 机：010-62770175　　邮　购：010-62786544
投稿与读者服务：010-62776969，c-service@tup.tsinghua.edu.cn
质量反馈：010-62772015，zhiliang@tup.tsinghua.edu.cn

印 装 者：三河市铭诚印务有限公司
经　　销：全国新华书店
开　　本：190mm×260mm　　印　张：22　　字　数：563千字
版　　次：2017年9月第1版　　印　次：2021年6月第4次印刷
定　　价：69.00元

产品编号：076840-01

推荐序

大数据是继石油之后，新兴的一种国家战略资源。大数据研究、开发和应用已经成为全球学术界、产业界的焦点。Hadoop、Spark等开源项目是目前大数据领域应用最广泛的技术和平台。熟练掌握Hadoop、Spark等是从事大数据研发和应用等从业人员必备的基本技能。

《Hadoop＋Spark生态系统操作与实战指南》正是在这样的技术背景下应运而生，能极好地满足广大大数据从业者的需求。本书以原理介绍为基础，以实战训练为目标，具体、深入地阐述了Hadoop及Spark的原生态中每一个组件的基本原理和应用方法；选择Apache和CDH两个主流Hadoop版本作为剖析实例，通过Java、Scala、客户端等开发案例，采用主流的离线项目和实时项目进行讲解。

作者根据自己多年在大数据行业的研发经验和个人体会，并结合大数据实际研发中需求和特点，认真整理其多年来编写的有关大数据研发的博文，精心组织和修订，最终编撰此著作，馈食读者。因此，该著作既是在大数据一线研发人员的知识结晶，而且还是有意进军大数据领域的从业人员的"良师益友"，确实是一本难得的大数据研发的参考资料。

黄永峰
清华大学电子工程系教授、博士生导师

随着大数据时代的到来，大数据技术在各行各业的应用越来越多，大数据相关技术的学习和使用者也越来越多。《Hadoop+Spark 生态系统操作与实战指南》从大数据爱好者和入门者的角度出发，以原理兼实战为主体思路展现 Hadoop 及 Spark 的原生态中每一个组件的操作方法，是一本有效的快速入门教程。

本书首先讲解了 Apache 和 CDH 两大 Hadoop 版本的集群搭建，并以此作为后续的开发平台；其次，讲解了 Hadoop+Spark 中原生态组件的原理，并使用 Java、Scala、客户端对组件进行实例操作，作为案例；最后，通过两个网页日志分析项目将 Hadoop 和 Spark 中的原生态组件整合在一起，作为项目架构。

余辉毕业于中国科学院大学，其研究方向为大数据与云计算，目前已拥有多年一线大数据开发经验。本书将理论与实践相结合，可作为相关技术教学和培训的参考资料。

肖俊

中国科学院大学人工智能技术学院教授、副院长

本书系统介绍了大数据相关知识，全书共有 12 章，论述了大数据的基本概念、大数据处理架构 Hadoop、分布式文件系统 HDFS、分布式数据库 HBase、NoSQL 数据库、云数据库、分布式并行编程模型 MapReduce、基于内存的分布式计算框架 Spark、最新的 ZooKeeper、Hive、Scala、Flume、Kafka 等技术。在 Hadoop、HDFS、HBase、MapReduce 和 Spark 等重要章节，都安排了实践操作，让读者更好地学习和掌握大数据关键技术。

本文作者余辉工程师，在大数据领域的实验室及公司工作多年，积累了丰富的实战经验。这本书理论结合实践，手把手教读者一步一步入门，避免了"纸上谈兵"，是大数据研究爱好者及从业人员的入门书籍。

本书可以作为高等院校计算机专业、信息管理等相关专业的大数据课程教材，也可供相关技术人员参考、学习、培训之用。

贺海武

中国科学院计算机网络信息中心（CNIC/CAS）百人计划研究员、

巴黎第 13 大学客座教授、里昂第 1 大学客座教授

市面上有许多讲解 Hadoop 或者 Spark 的书籍，但很难找到一本能带领大数据爱好者快速入门的书籍。本书作者余辉兼职于 Oracle OAEC 在线教育集团大数据讲师，他从一个讲师的角度写书，本书通过多维度讲解 Hadoop+Spark 原生态系统组件，在平台环境方面使用到 Apache 和 CDH 版本的 Hadoop 集群，在开发环境方面使用到 Eclipse+Java 和 IntelliJ IDEA+Scala，在项目环境方面使用到主流的离线日志分析和实时日志分析，让大数据爱好者可以快速认识大数据、熟悉大数据、操作大数据、运用大数据。本书详细讲解了 Hadoop+Spark 原生态组件的原理，通过 Java、Scala、客户端等开发案例并附上图片进行解说，让读者极易上手，本书非常适合作为一本大数据的快速入门教材。

方立勋

传智播客.黑马程序员高级副总裁

我与余辉的认识起源于清华大学，当时他在清华大学电子工程系担任软件工程师一职，通过和他多次交谈感觉此人思维缜密、善于总结且非常热爱技术。此书涵盖了余辉多年的一线开发经验和博文总结。

《Hadoop + Spark 生态系统操作与实战指南》总计 12 章。涵盖 Hadoop+Spark 原生态系统组件，对每一个组件原理和架构有着清晰的描述。通过两套主流开发环境 Eclipse+Java 和 IntelliJ IDEA+Scala 以及客户端分别对每一个组件进行了大量的案例操作，并配上大量案例截图，最后采用主流的离线项目和实时项目进行生态组件的融合。从多维度让读者对大数据快速认知、快速理解、快速上手、快速深入了解大数据行业，是一本非常适合大数据开发爱好者快速入门的书籍。

杨志云

搜狐视频技术总监

大数据在各行各业的应用越来越广,近几年"大数据"一词也非常火热,余辉的书《Hadoop + Spark 生态系统操作与实战指南》生逢其时。虽然现在世面上有不少关于大数据方面的书籍,但我还是想从本书的内容结构,及我与作者交往方面,对此书及此人做一个概要性的介绍。

此书最大的特点是理念、实战与项目的结合,能把各个知识点,以实战操作的方式连成线,再以项目的方式,把各知识模块连成面,点、线、面轮廓清晰、项目实用,能帮助读者快速理解大数据生态技术中的各种技术在实际应用中的作用。Hadoop 是大数据平台,它通过一系列的技术组成一个大数据生态技术圈,各种技术在这个生态中是干什么、原理是什么等在书中都有讲解。书中内容包括三大部分,12 章,从大数据生态平台起源讲起(第 1 章),实践环境搭建(第 2 章)、分布式存储与计算框架介绍(第 3 章)、平台协作套件(第 4 章)、Apache 原生的分布式计算框架详解(第 5 章)、分布式数据存储数据库(第 6、7 章)、利用函数式编程处理数据(第 8 章)、数据同步(第 9、10 章)、内存计算引擎架构(第 11 章)以及综合项目(第 12 章),内容丰富、案例真实、可操作性强,通过本书,读者能快速地理解 Hadoop 大数据技术生态中各种技术在实际项目中的应用。

关于此人,余辉是我通过 CSDN 博客找到他的,最开始我是阅读他的 CSDN 博文,从他的博文字里行间能感受到他的几种特质:专注、坚持、超强的执行力。因为 Oracle OAEC 人才产业中心此时正在开设大数据相关的课程,所以通过电话联系到他,经过一段时间的交流,最终成为 Oracle OAEC 人才产业基地的一名大数据兼职讲师,负责北京 Oracle OAEC 中心的大数据课程的教授。在教学过程中,得到学员的多次好评,以此基础,我建议他写一本关于这方面的书籍。我的逻辑是让他通过授课的方式,将多年在大数据一线的实际应用与项目,用通俗易懂的方式让学员理解;同时,自己也加深了理解;再通过写书的方式,能系统地将知识、经验、和自己的理解分享给别人。

刘 彰

Oracle OAEC 人才产业集团大数据学院与认证中心产品总监

ORACLE 认证高级讲师

前 言

近几年来，随着计算机和信息技术的迅猛发展和普及应用，行业应用系统的规模迅速扩大，行业应用所产生的数据呈爆炸性增长。大数据技术快速火热，大数据开发工程师更是供不应求。本书是一本 Hadoop+Spark 快速上手的书，从 Hadoop 生态系统和 Spark 生态系统全面原理解析和实战操作每一个组件，每一个知识点都讲得十分细致，让读者能够轻松地跨入大数据开发工程师的大门。

大数据工程师薪资

近几年大数据岗位尤其火热，大数据开发工程师供不应求，市面上大数据开发工程师起步就是 8 千元，1 年工作经验 1 万 2 千元，2 年工作经验 1 万 5 千元，3 年工作经验 2 万以上。根据每个人自身学习能力不同，有人 2 年就可以达到 2 万元以上。

下图是神州数码于 2017 年 6 月 6 日发布的一则招聘信息。

本书内容

全书共 12 章，分为 3 个部分，第 1 部分（第 1~7 章）讲解了 Hadoop 的原生态组件，包括 Hadoop、ZooKeeper、HBase、Hive 环境搭建与安装，以及如何对 MapReduce、HDFS、ZooKeeper、HBase、Hive 进行原理介绍和 Apache 版本环境下实战的操作。第 2 部分（第 8~11 章）讲解 Spark 的原生态组件，包括 Spark Core、Spark SQL、Spark Streaming、DataFrame，以及如何对 Scala、Spark API、Spark SQL、Spark Streaming、DataFrame 进行原理介绍和 CDH 版本环境下实战的操作，其中 Flume 和 Kafka 属于 Apache 顶级开源项目也放在本篇讲解。第 3 部分（第 12 章）讲解大数据项目，包络网页日志离线项目和实时项目，在 CDH 版本环境下通过两个项目将 Hadoop 和 Spark 原生态组件进行整合，一步步带领读者实战大数据项目。

本书特色

本书是一本 Hadoop + Spark 的快速入门书籍，以通俗易懂的方式介绍了 Hadoop + Spark 原生态组件的原理、实战操作以及集群搭建方面的知识。其中，Hadoop 原生态组件包括：MapReduce、HDFS、ZooKeeper、HBase、Hive；Spark 原生态组件包括：Spark Core、Spark SQL、Spark Streaming、Dataframe；同时包括 Apache 版本和 CDH5 版本的 Hadoop 集群搭建。本书的特点是：注重"实战"训练，强调知识系统性，关注内容实用性。

（1）本书从培训角度对读者简述 Hadoop + Spark 中常用组件的原理和实战操作，让读者快速了解组件原理和功能使用。

（2）每一个操作都配有实例代码或者图片来帮助理解，每一章的最后还有小节，以归纳总结本章的内容，帮助读者对 Hadoop + Spark 原生态系统有一个大的全局观。

（3）目前市面上关于 Hadoop 的书很多，关于 Spark 的书也很多，但是很少有对 Hadoop + Spark 结合进行讲解。本书首先讲解 Hadoop + Spark 原理，接着讲解 Hadoop + Spark 原生态组件的实例操作，最后结合大数据网站日志离线和实时两个项目融合 Hadoop+Spark 所有生态系统功能，使读者对本书有一个由浅入深且快速上手的过程。

本书适合读者

本书适合 Hadoop+Spark 的初学者，希望深入了解 Hadoop+Spark 安装部署、开发优化的大数据工程师，希望深入了解 Hadoop+Spark 管理、业务框架扩展的大数据架构师，以及任何对 Hadoop+Spark 相关技术感兴趣的读者。

本书代码、软件、文档下载

本书代码、软件、文档下载地址(注意数字和字母大小写)如下:
http://pan.baidu.com/s/1cCi0k2

如果下载有问题,请联系电子邮箱 booksaga@163.com,邮件主题为"Hadoop+Spark 生态系统与实战指南"。

本书作者

余辉,中国科学院大学硕士研究生毕业,研究方向为云计算和大数据。现供职于某上市公司技术经理,并在 Oracle OAEC 人才产业集团大数据学院(http://www.oracleoaec.com.cn/)担任大数据讲师。曾在清华大学电子工程系 NGNLab 研究室(http://ngn.ee.tsinghua.edu.cn/)担任软件工程师。

已发表两篇大数据论文:《微博舆情的 Hadoop 存储和管理平台设计与实现》和《跨媒体多源网络舆情分析系统设计与实现》。

博客:http://blog.csdn.net/silentwolfyh
微博:http://weibo.com/u/3195228233
电子邮箱:yuhuiqh2009@163.com

致谢

赶在儿子 1 岁生日之际,赶在我告别 30 岁之际,我撰写《Hadoop+Spark 生态系统操作与实战指南》一书,作为我儿子的生日礼物。感谢父母提供了良好的生活环境,感谢舅舅、舅妈提供了良好的学习平台,感谢我的老婆、姐姐、姐夫在生活上对我的支持和奉献。最后,感谢清华工作和学习的那些时光,清华六年,我学会了生存技能、找到了研究方向、培养了生活习惯。

余 辉
2017 年 7 月

目 录

第1章 Hadoop 概述 ... 1
- 1.1 Hadoop 简介 ... 1
- 1.2 Hadoop 版本和生态系统 ... 3
- 1.3 MapReduce 简介 ... 7
- 1.4 HDFS 简介 ... 8
- 1.5 Eclipse+Java 开发环境搭建 ... 10
  - 1.5.1 Java 安装 ... 10
  - 1.5.2 Maven 安装 ... 11
  - 1.5.3 Eclipse 安装和配置 ... 12
  - 1.5.4 Eclipse 创建 Maven 项目 ... 16
  - 1.5.5 Eclipse 其余配置 ... 19
- 1.6 小结 ... 21

第2章 Hadoop 集群搭建 ... 22
- 2.1 虚拟机简介 ... 22
- 2.2 虚拟机配置 ... 24
- 2.3 Linux 系统设置 ... 31
- 2.4 Apache 版本 Hadoop 集群搭建 ... 36
- 2.5 CDH 版本 Hadoop 集群搭建 ... 44
  - 2.5.1 安装前期准备 ... 44
  - 2.5.2 Cloudera Manager 安装 ... 45
  - 2.5.3 CDH 安装 ... 46
- 2.6 小结 ... 55

第3章 Hadoop 基础与原理 ... 56
- 3.1 MapReduce 原理介绍 ... 56
  - 3.1.1 MapReduce 的框架介绍 ... 56
  - 3.1.2 MapReduce 的执行步骤 ... 58
- 3.2 HDFS 原理介绍 ... 59
  - 3.2.1 HDFS 是什么 ... 59
  - 3.2.2 HDFS 架构介绍 ... 59
- 3.3 HDFS 实战 ... 62

    3.3.1 HDFS 客户端的操作 62

    3.3.2 Java 操作 HDFS 65

  3.4 YARN 原理介绍 69

  3.5 小结 71

第 4 章 ZooKeeper 实战 72

  4.1 ZooKeeper 原理介绍 72

    4.1.1 ZooKeeper 基本概念 72

    4.1.2 ZooKeeper 工作原理 73

    4.1.3 ZooKeeper 工作流程 76

  4.2 ZooKeeper 安装 78

  4.3 ZooKeeper 实战 80

    4.3.1 ZooKeeper 客户端的操作 80

    4.3.2 Java 操作 ZooKeeper 81

    4.3.3 Scala 操作 ZooKeeper 85

  4.4 小结 87

第 5 章 MapReduce 实战 88

  5.1 前期准备 88

  5.2 查看 YARN 上的任务 95

  5.3 加载配置文件 95

  5.4 MapReduce 实战 96

  5.5 小结 121

第 6 章 HBase 实战 122

  6.1 HBase 简介及架构 122

  6.2 HBase 安装 127

  6.3 HBase 实战 129

    6.3.1 HBase 客户端的操作 129

    6.3.2 Java 操作 HBase 132

    6.3.3 Scala 操作 HBase 136

  6.4 小结 140

第 7 章 Hive 实战 141

  7.1 Hive 介绍和架构 141

  7.2 Hive 数据类型和表结构 143

  7.3 Hive 分区、桶与倾斜 144

  7.4 Hive 安装 146

  7.5 Hive 实战 148

		7.5.1 Hive 客户端的操作 148
		7.5.2 Hive 常用命令 154
		7.5.3 Java 操作 Hive 155
	7.6	小结 161

第 8 章 Scala 实战 162

- 8.1 Scala 简介与安装 162
- 8.2 IntelliJ IDEA 开发环境搭建 164
  - 8.2.1 IntelliJ IDEA 简介 164
  - 8.2.2 IntelliJ IDEA 安装 164
  - 8.2.3 软件配置 166
- 8.3 IntelliJ IDEA 建立 Maven 项目 171
- 8.4 基础语法 176
- 8.5 函数 179
- 8.6 控制语句 181
- 8.7 函数式编程 184
- 8.8 模式匹配 189
- 8.9 类和对象 191
- 8.10 Scala 异常处理 194
- 8.11 Trait（特征） 195
- 8.12 Scala 文件 I/O 196
- 8.13 作业 198
  - 8.13.1 九九乘法表 198
  - 8.13.2 冒泡排序 199
  - 8.13.3 设计模式 Command 200
  - 8.13.4 集合对称判断 202
  - 8.13.5 综合题 204
- 8.14 小结 206

第 9 章 Flume 实战 207

- 9.1 Flume 概述 207
- 9.2 Flume 的结构 208
- 9.3 Flume 安装 211
- 9.4 Flume 实战 212
- 9.5 小结 214

第 10 章 Kafka 实战 215

- 10.1 Kafka 概述 215

- 10.1.1 简介 215
- 10.1.2 使用场景 217
- 10.2 Kafka 设计原理 218
- 10.3 Kafka 主要配置 222
- 10.4 Kafka 客户端操作 224
- 10.5 Java 操作 Kafka 226
  - 10.5.1 生产者 226
  - 10.5.2 消费者 228
- 10.6 Flume 连接 Kafka 229
- 10.7 小结 233

第 11 章 Spark 实战 234

- 11.1 Spark 概述 234
- 11.2 Spark 基本概念 234
- 11.3 Spark 算子实战及功能描述 238
  - 11.3.1 Value 型 Transformation 算子 238
  - 11.3.2 Key-Value 型 Transformation 算子 242
  - 11.3.3 Actions 算子 245
- 11.4 Spark Streaming 实战 248
- 11.5 Spark SQL 和 DataFrame 实战 253
- 11.6 小结 266

第 12 章 大数据网站日志分析项目 267

- 12.1 项目介绍 267
- 12.2 网站离线项目 267
  - 12.2.1 业务框架图 267
  - 12.2.2 子服务"趋势分析"详解 268
  - 12.2.3 表格的设计 272
  - 12.2.4 提前准备 274
  - 12.2.5 项目步骤 287
- 12.3 网站实时项目 297
  - 12.3.1 业务框架图 297
  - 12.3.2 子服务"当前在线"详解 297
  - 12.3.3 表格的设计 302
  - 12.3.4 提前准备 304
  - 12.3.5 项目步骤 327
- 12.4 小结 337

第 1 章

◀ Hadoop 概述 ▶

1.1 Hadoop 简介

1. Hadoop 的由来

Hadoop 是 Doug Cutting（Apache Lucene 创始人）开发的、使用广泛的文本搜索库。Hadoop 起源于 Apache Nutch，后者是一个开源的网络搜索引擎，本身也是 Lucene 项目的一部分。

2. Hadoop 名字的起源

Hadoop 这个名字不是一个缩写，它是一个虚构的名字。该项目的创建者 Doug Cutting 如此解释 Hadoop 的得名："这个名字是我孩子给一头吃饱了的棕黄色大象命名的。我的命名标准就是简短、容易发音和拼写，没有太多的意义，并且不会被用于别处。小孩子是这方面的高手。Googol 就是由小孩命名的。"（Google 来源于 Googol 一词。GooGol 指的是 10 的 100 次幂（方），代表互联网上的海量资源。公司创建之初，肖恩·安德森在搜索该名字是否已经被注册时，将 Googol 误打成了 Google。）

Hadoop 及其子项目和后继模块所使用的名字往往也与其功能不相关，经常用一头大象或其他动物主题（例如：Pig）。较小的各个组成部分给予更多描述性（因此也更俗）的名称。这是一个很好的原则，因为它意味着可以大致从其名字猜测其功能，例如，jobtracker 的任务就是跟踪 MapReduce 作业。

从头开始构建一个网络搜索引擎是一个雄心勃勃的目标，不只是要编写一个复杂的、能够抓取和索引网站的软件，还需要面临着没有专业运行团队支持运行它的挑战，因为它有那么多独立部件。同样昂贵的还有：据 Mike Cafarella 和 Doug Cutting 估计，一个支持此 10 亿页的索引，需要价值约 50 万美元的硬件投入，每月运行费用还需要 3 万美元。不过，他们相信这是一个有价值的目标，因为这会开放并最终使搜索引擎算法普及化。

Nutch 项目开始于 2002 年，一个可工作的抓取工具和搜索系统很快浮出水面。但他们意识到，他们的架构将无法扩展到拥有数十亿网页的网络。在 2003 年发表的一篇描述 Google 分布式文件系统（简称 GFS）的论文为他们提供了及时的帮助，文中称 Google 正在使用此文件系统。GFS 或类似的东西，可以解决他们在网络抓取和索引过程中产生的大量的文件的存储需求。具体而言，GFS 会省掉管理所花的时间，如管理存储节点。在 2004 年，他们开始写一个开放源码的应用，即 Nutch 的分布式文件系统（NDFS）。

2004 年，Google 发表了论文，向全世界介绍了 MapReduce。2005 年初，Nutch 的开发者在 Nutch 上有了一个可工作的 MapReduce 应用，到当年年中，所有主要的 Nutch 算法被移植到使用 MapReduce 和 NDFS 来运行。

Nutch 中的 NDFS 和 MapReduce 实现的应用远不只是搜索领域，在 2006 年 2 月，他们从 Nutch 转移出来成为一个独立的 Lucene 子项目，称为 Hadoop。大约在同一时间，Doug Cutting 加入雅虎，Yahoo 提供一个专门的团队和资源将 Hadoop 发展成一个可在网络上运行的系统（见后文的补充材料）。在 2008 年 2 月，雅虎宣布其搜索引擎产品部署在一个拥有 1 万个内核的 Hadoop 集群上。

2008 年 1 月，Hadoop 已成为 Apache 顶级项目，证明它是成功的，是一个多样化、活跃的社区。通过这次机会，Hadoop 成功地被雅虎之外的很多公司应用，如 Last.fm、Facebook 和《纽约时报》。一些应用在 Hadoop 维基有介绍，Hadoop 维基的网址为 http://wiki.apache.org/hadoop/PoweredBy。

有一个良好的宣传范例，《纽约时报》使用亚马逊的 EC2 云计算将 4 TB 的报纸扫描文档压缩，转换为用于 Web 的 PDF 文件。这个过程历时不到 24 小时，使用 100 台机器运行，如果不结合亚马逊的按小时付费的模式（即允许《纽约时报》在很短的一段时间内访问大量机器）和 Hadoop 易于使用的并行程序设计模型，该项目很可能不会这么快开始启动。

2008 年 4 月，Hadoop 打破世界纪录，成为最快排序 1 TB 数据的系统，运行在一个 910 节点的集群，Hadoop 在 209 秒内排序了 1 TB 的数据（还不到三分半钟），击败了前一年的 297 秒冠军。同年 11 月，谷歌在报告中声称，它的 MapReduce 实现执行 1 TB 数据的排序只用了 68 秒。在 2009 年 5 月，有报道宣称 Yahoo 的团队使用 Hadoop 对 1 TB 的数据进行排序只花了 62 秒时间。

构建互联网规模的搜索引擎需要大量的数据，因此需要大量的机器来进行处理。Yahoo！Search 包括四个主要组成部分：Crawler，从因特网下载网页；WebMap，构建一个网络地图；Indexer，为最佳页面构建一个反向索引；Runtime（运行时），回答用户的查询。WebMap 是一幅图，大约包括一万亿条边（每条代表一个网络链接）和一千亿个节点（每个节点代表不同的网址）。创建和分析此类大图需要大量计算机运行若干天。在 2005 年初，WebMap 所用的基础设施名为 Dreadnaught，需要重新设计以适应更多节点的需求。Dreadnaught 成功地从 20 个节点扩展到 600 个，但还需要一个完全重新的设计，以进一步扩大。Dreadnaught 与 MapReduce 有许多相似的地方，但灵活性更强，结构更少。具体说来，Dreadnaught 作业可以将输出发送到此作业下一阶段中的每一个分段（fragment），但排序是在库函数中完成的。在实际情形中，大多数 WebMap 阶段都是成对存在的，对应于 MapReduce。因此，WebMap 应用并不需要为了适应 MapReduce 而进行大量重构。

Eric Baldeschwieler（Eric14）组建了一个小团队，他们开始设计并原型化一个新的框架（原型为 GFS 和 MapReduce，用 C++语言编写），打算用它来替换 Dreadnaught。尽管当务之急是需要一个 WebMap 新框架，但显然，标准化对于整个 Yahoo! Search 平台至关重要，并且通过使这个框架泛化，足以支持其他用户，这样他们才能够充分运用对整个平台的投资。

与此同时，雅虎在关注 Hadoop（当时还是 Nutch 的一部分）及其进展情况。2006 年 1

月,雅虎聘请了 Doug Cutting,一个月后,决定放弃自己的原型,转而使用 Hadoop。相较于雅虎自己的原型和设计,Hadoop 的优势在于它已经在 20 个节点上实际应用过。这样一来,雅虎便能在两个月内搭建一个研究集群,并着手帮助真正的客户使用这个新的框架,速度比原来预计的快许多。另一个明显的优点是 Hadoop 已经开源,较容易(虽然远没有那么容易!)从雅虎法务部门获得许可在开源方面进行工作。因此,雅虎在 2006 年初设立了一个 200 个节点的研究集群,他们将 WebMap 的计划暂时搁置,转而为研究用户支持和发展 Hadoop。

3. Hadoop 大事记

2004 年,最初的版本(现在称为 HDFS 和 MapReduce)由 Doug Cutting 和 Mike Cafarella 开始实施。

2005 年 12 月,Nutch 移植到新的框架,Hadoop 在 20 个节点上稳定运行。

2006 年 1 月,Doug Cutting 加入雅虎。

2006 年 2 月,Apache Hadoop 项目正式启动以支持 MapReduce 和 HDFS 的独立发展。

2006 年 2 月,雅虎的网格计算团队采用 Hadoop。

2006 年 4 月,标准排序(10 GB 每个节点)在 188 个节点上运行 47.9 个小时。

2006 年 5 月,雅虎建立了一个 300 个节点的 Hadoop 研究集群。

2006 年 5 月,标准排序在 500 个节点上运行 42 个小时(硬件配置比 4 月的更好)。

2006 年 11 月,研究集群增加到 600 个节点。

2006 年 12 月,标准排序在 20 个节点上运行 1.8 个小时,100 个节点 3.3 小时,500 个节点 5.2 小时,900 个节点 7.8 个小时。

2007 年 1 月,研究集群到达 900 个节点。

2007 年 4 月,研究集群达到两个 1000 个节点的集群。

2008 年 4 月,赢得世界最快 1 TB 数据排序在 900 个节点上用时 209 秒。

2008 年 10 月,研究集群每天装载 10 TB 的数据。

2009 年 3 月,17 个集群总共 24 000 台机器。

2009 年 4 月,赢得每分钟排序,59 秒内排序 500 GB(在 1400 个节点上)和 173 分钟内排序 100 TB 数据(在 3400 个节点上)。

1.2 Hadoop 版本和生态系统

1. Hadoop 版本的优缺点

目前市面上 Hadoop 版本主要有两种:Apache 版本和 CDH 版本。

(1)Aapche 版本的 Hadoop

官网:http://hadoop.apache.org/

Aapche Hadoop 优势:对硬件要求低。

Aapche Hadoop 劣势:搭建烦琐,维护烦琐,升级烦琐,添加组件烦琐。

Apache 版本 Hadoop 集群中 YARN 的界面如图 1-1 所示，HDFS 的界面图 1-2 所示。

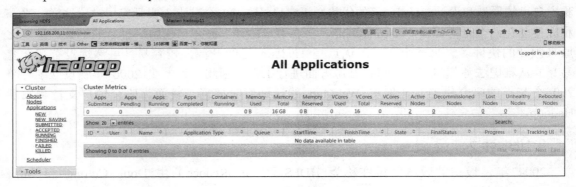

图 1-1 YARN 的界面

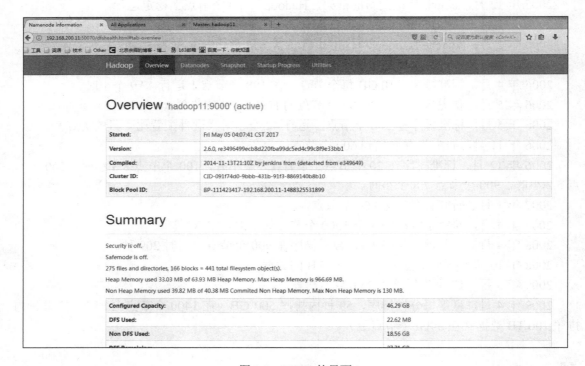

图 1-2 HDFS 的界面

（2）CDH 版本的 Hadoop

官网：https://www.cloudera.com/

CDH 优势：搭建方便，维护较为容易，升级以及迁移容易，添加组件容易。

CDH 缺点：对硬件要求高。

Cloudera Manager 是一个管理 CDH 的端到端的应用。主要作用包括：管理、监控、诊断、集成。

CDH 的 Hadoop 版本集群中 CDH 管理界面如图 1-3 所示。

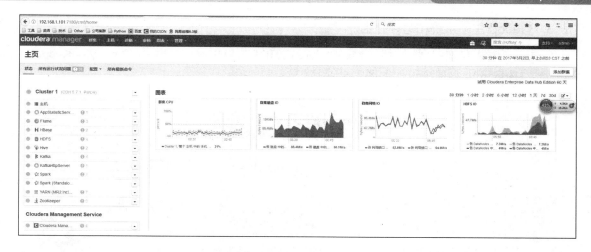

图 1-3　CDH 管理界面

2. CDH 架构

CDH 架构如图 1-4 所示。

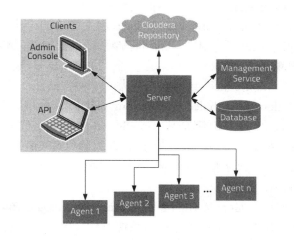

图 1-4　CDH 架构

（1）Server

管理控制台服务器和应用程序逻辑。

负责软件安装、配置，启动和停止服务。

管理服务运行的集群。

（2）Agent

安装在每台主机上。

负责启动和停止进程，配置、监控主机。

（3）Management Service

由一组角色组成的服务，执行各种监视、报警和报告功能。

3. Hadoop 生态系统和组件介绍

Hadoop 生态组件主要包括：MapReduce、HDFS、HBase、Hive、Pig、ZooKeeper、Mahout。Hadoop 生态系统组件如图 1-5 所示。

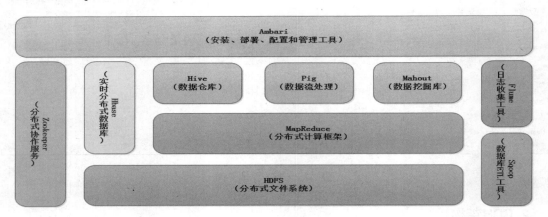

图 1-5　Hadoop 生态系统组件

（1）MapReduce：MapReduce 是使用集群的并行、分布式算法处理大数据集的可编程模型。Apache MapReduce 是从 Google MapReduce 派生而来的：在大型集群中简化数据处理。当前的 Apache MapReduce 版本基于 Apache YARN 框架构建。YARN 是 "Yet-Another-Resource-Negotiator" 的缩写。YARN 可以运行非 MapReduce 模型的应用。YARN 是 Apache Hadoop 想要超越 MapReduce 数据处理能力的一种尝试。

（2）HDFS：The Hadoop Distributed File System（HDFS）提供跨多个机器存储大型文件的一种解决方案。Hadoop 和 HDFS 都是从 Google File System（GFS）中派生的。Hadoop 2.0.0 之前，NameNode 是 HDFS 集群的一个单点故障（SPOF）。利用 ZooKeeper、HDFS 高可用性特性解决了这个问题，提供选项来运行两个重复的 NameNodes，在同一个集群中，同一个 Active/Passive 配置。

（3）HBase：灵感来源于 Google BigTable。HBase 是 Google Bigtable 的开源实现，类似 Google Bigtable 利用 GFS 作为其文件存储系统，HBase 利用 Hadoop HDFS 作为其文件存储系统；Google 运行 MapReduce 来处理 Bigtable 中的海量数据，HBase 同样利用 Hadoop MapReduce 来处理 HBase 中的海量数据；Google Bigtable 利用 Chubby 作为协同服务，HBase 利用 ZooKeeper 作为对应。

（4）Hive：Facebook 开发的数据仓库基础设施。数据汇总、查询和分析。Hive 提供类似 SQL 的语言（不兼容 SQL92）：HiveQL。

（5）Pig：Pig 提供一个引擎在 Hadoop 并行执行数据流。Pig 包含一个语言：Pig Latin，用来表达这些数据流。Pig Latin 包括大量的传统数据操作（join、sort、filter 等），也可以让用户开发他们自己的函数，用来查看、处理和编写数据。Pig 在 Hadoop 上运行，在 Hadoop 分布式文件系统（HDFS）和 Hadoop 处理系统 MapReduce 中都有使用。Pig 使用 MapReduce 来执行所有的数据处理，编译 Pig Latin 脚本，用户可以编写一个系列，一个或者多个的 MapReduce 作业，然后执行。Pig Latin 看起来跟大多数编程语言都不一样，没有

if 状态和 for 循环。

（6）ZooKeeper：ZooKeeper 是 Hadoop 的正式子项目，它是一个针对大型分布式系统的可靠协调系统，提供的功能包括：配置维护、名字服务、分布式同步、组服务等。ZooKeeper 的目标就是封装好复杂易出错的关键服务，将简单易用的接口和性能高效、功能稳定的系统提供给用户。ZooKeeper 是 Google 的 Chubby 一个开源的实现，是高效和可靠的协同工作系统。ZooKeeper 能够用来执行 leader 选举、配置信息维护等。在一个分布式的环境中，我们需要一个 Master 实例用来存储一些配置信息，确保文件写入的一致性等。

（7）Mahout：基于 MapReduce 的机器学习库和数学库。

1.3 MapReduce 简介

1. MapReduce 是什么

MapReduce 是面向大数据并行处理的计算模型、框架和平台，它隐含了以下三层含义：

（1）MapReduce 是一个基于集群的高性能并行计算平台（Cluster Infrastructure）。它允许用市场上普通的商用服务器构成一个包含数十、数百至数千个节点的分布和并行计算集群。

（2）MapReduce 是一个并行计算与运行软件框架（Software Framework）。它提供了一个庞大但设计精良的并行计算软件框架，能自动完成计算任务的并行化处理，自动划分计算数据和计算任务，在集群节点上自动分配和执行任务以及收集计算结果，将数据分布存储、数据通信、容错处理等并行计算涉及的很多系统底层的复杂细节交由系统负责处理，大大减少了软件开发人员的负担。

（3）MapReduce 是一个并行程序设计模型与方法（Programming Model & Methodology）。它借助于函数式程序设计语言 Lisp 的设计思想，提供了一种简便的并行程序设计方法，用 Map 和 Reduce 两个函数编程实现基本的并行计算任务，提供了抽象的操作和并行编程接口，以简单方便地完成大规模数据的编程和计算处理。

2. MapReduce 的由来

MapReduce 最早是由 Google 公司研究提出的一种面向大规模数据处理的并行计算模型和方法。Google 公司设计 MapReduce 的初衷主要是为了解决其搜索引擎中大规模网页数据的并行化处理。Google 公司发明了 MapReduce 之后，首先用其重新改写了搜索引擎中的 Web 文档索引处理系统。但由于 MapReduce 可以普遍应用于很多大规模数据的计算问题，因此自发明 MapReduce 以后，Google 公司内部进一步将其广泛应用于很多大规模数据处理问题。到目前为止，Google 公司内有上万个各种不同的算法问题和程序都使用 MapReduce 进行处理。

2003 年和 2004 年，Google 公司在国际会议上分别发表了两篇关于 Google 分布式文件系统和 MapReduce 的论文，公布了 Google 的 GFS 和 MapReduce 的基本原理和主要设计思想。2004 年，开源项目 Lucene（搜索索引程序库）和 Nutch（搜索引擎）的创始人 Doug Cutting

发现 MapReduce 正是其所需要的解决大规模 Web 数据处理的重要技术，因而模仿 Google MapReduce，基于 Java 设计开发了一个称为 Hadoop 的开源 MapReduce 并行计算框架和系统。自此，Hadoop 成为 Apache 开源组织下最重要的项目，自其推出后很快得到了全球学术界和工业界的普遍关注，并得到推广和普及应用。

MapReduce 的推出给大数据并行处理带来了巨大的革命性影响，使其已经成为事实上的大数据处理的工业标准。尽管 MapReduce 还有很多局限性，但人们普遍公认，MapReduce 是到目前为止更为成功、更广为接受和更易于使用的大数据并行处理技术。MapReduce 的发展普及和带来的巨大影响远远超出了发明者和开源社区当初的意料，以至于马里兰大学教授、2010 年出版的 *Data-Intensive Text Processing with MapReduce* 一书的作者 Jimmy Lin 在书中提出：MapReduce 改变了我们组织大规模计算的方式，它代表了第一个有别于冯·诺依曼结构的计算模型，是在集群规模而非单个机器上组织大规模计算的、新的抽象模型上的第一个重大突破，是到目前为止所见到的最为成功的基于大规模计算资源的计算模型。

3. MapReduce 设计目标

MapReduce 是一种可用于数据处理的编程框架。MapReduce 采用"分而治之"的思想，把对大规模数据集的操作，分发给一个主节点管理下的各个分节点共同完成，然后通过整合各个节点的中间结果，得到最终结果。简单地说，MapReduce 就是"任务的分解与结果的汇总"。

在分布式计算中，MapReduce 框架负责处理并行编程中分布式存储、工作调度、负载均衡、容错均衡、容错处理以及网络通信等复杂问题，把处理过程高度抽象为两个函数：map 和 reduce，map 负责把任务分解成多个任务，reduce 负责把分解后多任务处理的结果汇总起来。

4. MapReduce 适用场景

用 MapReduce 来处理的数据集（或任务）必须具备这样的特点：待处理的数据集可以分解成许多小的数据集，而且每一个小数据集都可以完全并行地进行处理。

1.4 HDFS 简介

1. HDFS 是什么

Hadoop 分布式文件系统（HDFS）被设计成适合运行在通用硬件（commodity hardware）上的分布式文件系统。它和现有的分布式文件系统有很多共同点。但同时，它和其他的分布式文件系统的区别也是很明显的。HDFS 是一个高度容错性的系统，适合部署在廉价的机器上。HDFS 能提供高吞吐量的数据访问，非常适合大规模数据集上的应用。HDFS 放宽了一部分 POSIX 约束，来实现流式读取文件系统数据的目的。HDFS 在最开始是作为 Apache Nutch 搜索引擎项目的基础架构而开发的。目前 HDFS 是 Apache Hadoop Core 项目的一部分。

2. HDFS 的由来

我们知道 HDFS 源于 Google 发布的 GFS 论文。HDFS 是 Hadoop Distribute File System 的简称，是 Hadoop 的一个分布式文件系统。

3. HDFS 设计目标

- 大文件存储：支持 TB、PB 级的数据量。
- 高容错：运行在商业硬件上，而商业硬件并不可靠。
- 高吞吐量：为大量数据访问的应用提供高吞吐量支持。

4. HDFS 适用场景与不适用场景

HDFS 适用场景与不适用场景如表 1-1 所示。

表 1-1 HDFS 适用场景与不适用场景

适用场景	不适用场景
大文件访问	存储大量小文件
流式数据访问	随机读取，低延迟读取

5. HDFS 的基本概念

（1）数据块（block）

大文件会被分割成多个 block 进行存储，block 大小默认为 128MB。小于一个块大小的文件不会占据整个块的空间。

每一个 block 会在多个 DataNode 上存储多份副本，默认是 3 份。

比如：1MB 的文件有 1 个 block 块，128MB 的文件有 1 个 block 块，129MB 的文件有 2 个 block 块。

（2）NameNode

记录文件系统的元数据，单一主元数据服务器，其中包括每个文件、文件位置以及这些文件所在的 DataNode 内的所有数据块的内存映射。维护文件系统树及整棵树内所有的文件和目录。

（3）DataNode

负责存储和检索数据块，它受客户端和 NameNode 调度，并且它会定期向 NameNode 发送本节点上所存储的块列表，这就是为什么 NameNode 并不是永久保存各个节点块信息的原因了。

1.5 Eclipse+Java 开发环境搭建

1.5.1 Java 安装

1. Java 安装步骤

本书中 Java 版本为 java1.8.0_77，安装步骤如图 1-6 至图 1-9 所示，不再详细说明，读者注意一下安装的版本即可。

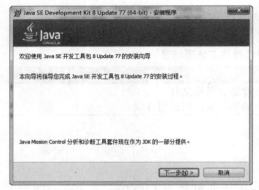

图 1-6　安装步骤一

图 1-7　安装步骤二

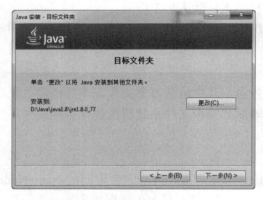

图 1-8　安装步骤三

图 1-9　安装步骤四

2. Java 环境变量及测试

配置 Java 环境变量，步骤为：控制面板→系统和安全→系统→高级系统设置。Java 环境变量设置如图 1-10 所示。

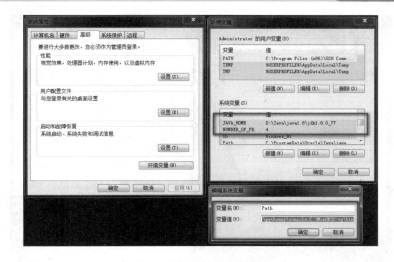

图 1-10　Java 环境变量设置

测试 Java 安装是否成功，在命令行窗口运行 java -version 命令。Java 测试如图 1-11 所示。

图 1-11　Java 测试

1.5.2　Maven 安装

1. Maven 下载

本书中采用的是 Maven3.3.3 版本，下载之后，把文件 apache-maven-3.3.3-bin.tar.gz 解压放入 D:\Java 目录中。

2. Maven 配置

Maven 下载后，需要配置环境变量，将 Maven 解压后的 bin 目录配置到环境变量中，Maven 环境变量设置如图 1-12 所示，这里用的是 3.3.3 版本，并安装在 D 盘根目录下。

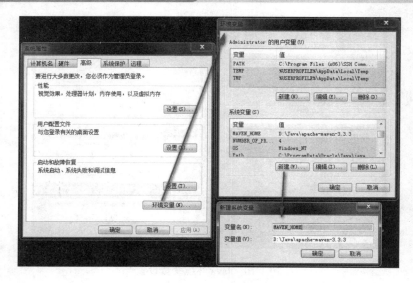

图 1-12　Maven 环境变量设置

3. Maven 测试

可打开命令行，输入命令 mvn -v 进行测试。Maven 测试如图 1-13 所示（出现版本信息，即表示配置成功）。

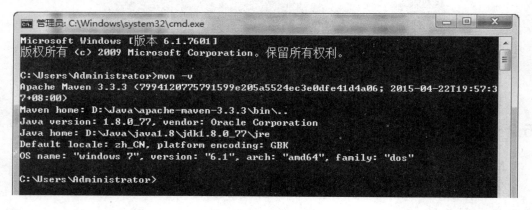

图 1-13　Maven 测试

1.5.3　Eclipse 安装和配置

本书中 Eclipse 版本为 eclipse4.4.2，下载解压之后放入 D:\Java 目录下面，且创建快捷键到桌面。Eclipse 解压结果如图 1-14 所示，Eclipse 创建启动快捷方式如图 1-15 所示。

图 1-14　Eclipse 解压结果

第 1 章　Hadoop 概述

图 1-15　Eclipse 创建启动快捷方式

创建 Eclipse 的 workspace 目录 D:\Java\eclipseworkspace，并设置 Eclipse 的工作空间 eclipseworkspace，如图 1-16 所示。

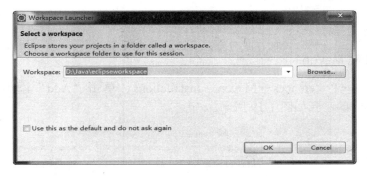

图 1-16　设置 Eclipse 的 eclipseworkspace

1. Eclipse 加载本地 Java

Eclipse 添加 Java 步骤如图 1-17 所示。

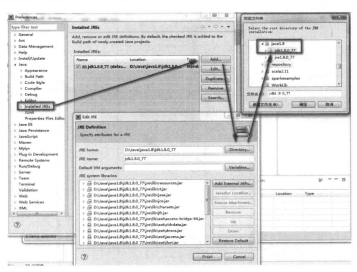

图 1-17　Eclipse 添加 Java

13

2. Eclipse 运行测试

建立测试类运行如图 1-18 所示。

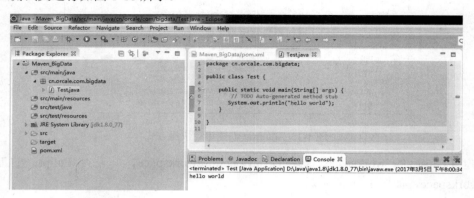

图 1-18　Eclipse 建立测试类运行

3. Eclipse 加载 Maven

选择 Window→Preferences→Maven→Installations，单击"Add"添加 Maven 目录。Eclipse 加载 Maven 的步骤如图 1-19 所示。

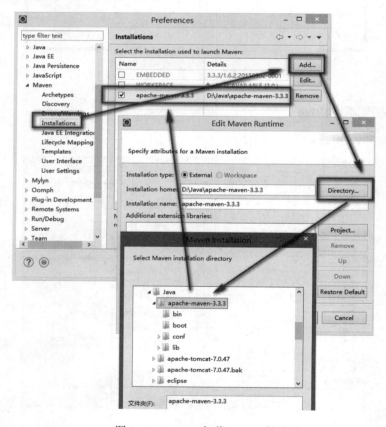

图 1-19　Eclipse 加载 Maven 的步骤

4. Eclipse 中 maven 配置

默认的 Maven 仓库位置为：C:\Users\Administrator\.m2（其中 Administrator 为当前系统用户账号）。Eclipse 设置 Maven 仓库的位置如图 1-20 所示。

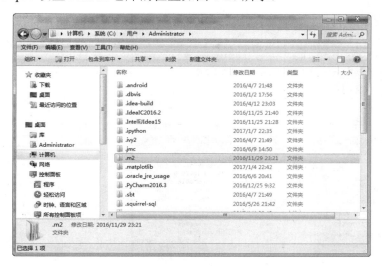

图 1-20　Eclipse 设置 Maven 仓库的位置

Maven 仓库位置的配置文件是 D:\Java\apache-maven-3.3.3\conf\settings.xml。settings.xml 设置内容如图 1-21 所示，Eclipse 加载 settings.xml 步骤如图 1-22 所示。

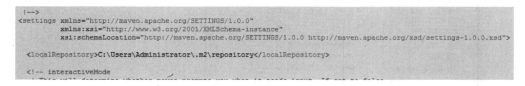

图 1-21　settings.xml 设置内容

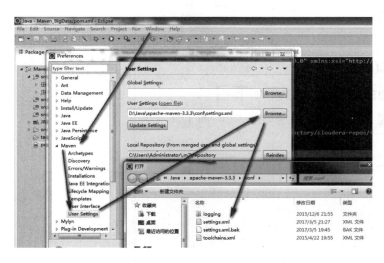

图 1-22　Eclipse 加载 settings.xml 步骤

5. Eclipse 中 Maven 测试

这是 Eclipse 中的 pom.xml 文件，加入下列内容保存之后，Eclipse 就自动下载依赖包。pom.xml 自动下载的包如图 1-23 所示。

```
<dependencies>
<dependency>
<groupId>org.apache.kafka</groupId>
<artifactId>kafka-clients</artifactId>
<version>0.9.0.1</version>
</dependency>
</dependencies>
```

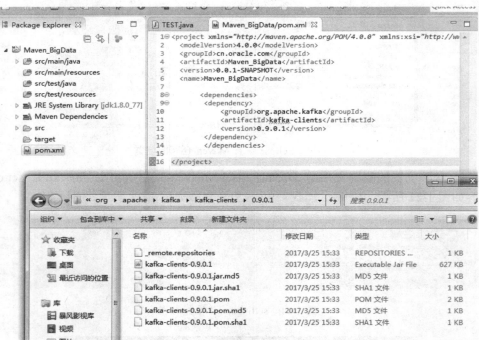

图 1-23　pom.xml 自动下载的包

1.5.4　Eclipse 创建 Maven 项目

Eclipse 创建 Maven 项目步骤如图 1-24~图 1-26 所示。

第 1 章　Hadoop 概述

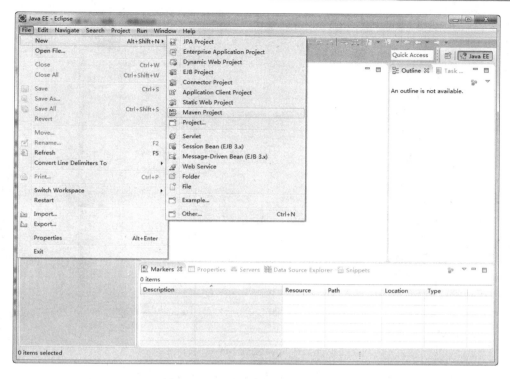

图 1-24　Eclipse 创建 Maven 项目一

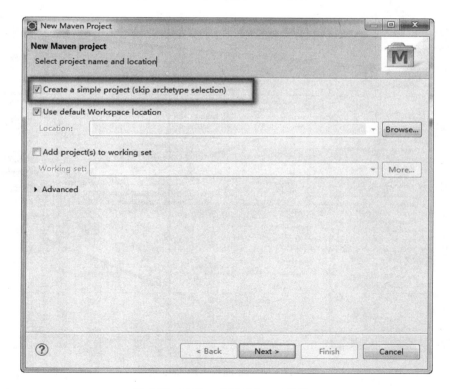

图 1-25　Eclipse 创建 Maven 项目二

17

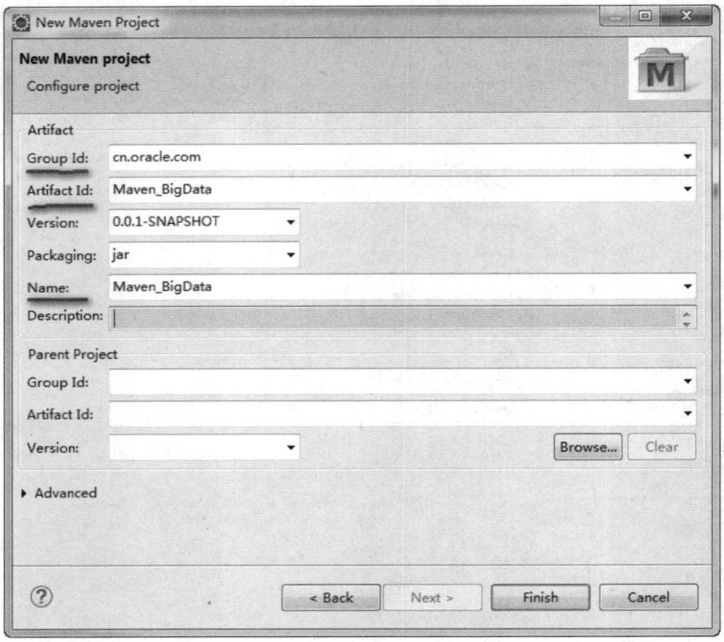

图 1-26　Eclipse 创建 Maven 项目三

单击 Finish 按钮完成。Maven 加载本地 Java 步骤如图 1-27 所示。

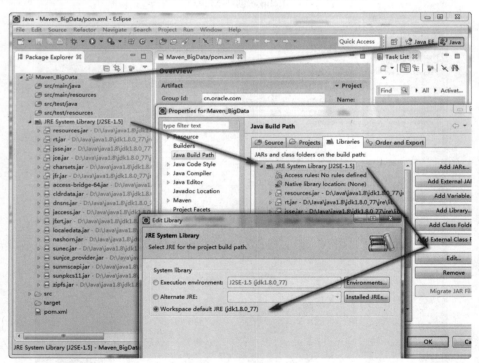

图 1-27　Maven 加载本地 Java 步骤

项目建立完后 Eclipse 的整体展示如图 1-28 所示。

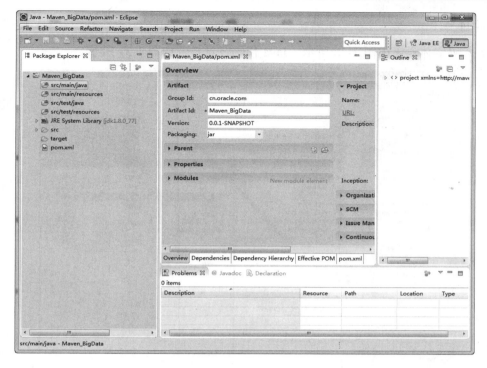

图 1-28 项目建立完后 Eclipse 的整体展示

1.5.5 Eclipse 其余配置

1. Eclipse 中 Runnable JAR file 的历史记录

用 Eclipse 打包 jar 的时候，需要指定一个 main 函数。需要先运行一下 main 函数，Eclipse 的 Runnable JAR File Specification 下的 Launch configuration 下拉列表才会有记录。如果想要删除下拉列表里的历史记录，只需要进入到要打包的这个工程所在的目录。

- Linux：/.metadata/.plugins/org.eclipse.debug.core/.launches
- Windows：E:\Java\eclipseworkspace\.metadata\.plugins\org.eclipse.debug.core\.launches

2. 主题

Eclipse 主题设置如图 1-29 所示。

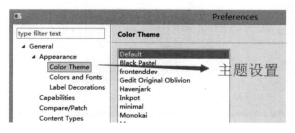

图 1-29 Eclipse 主题设置

3. 字体

Java 代码字体设置如图 1-30 所示，Java 代码背景设置如图 1-31 所示。

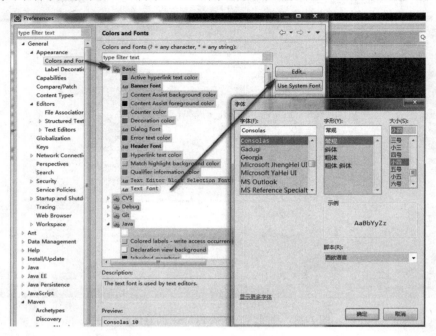

图 1-30　Java 代码字体设置

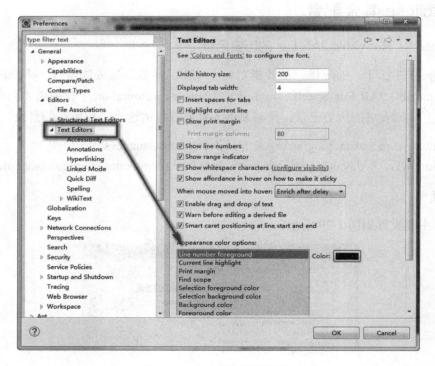

图 1-31　Java 代码背景设置

4. Debug 当前行颜色

Debug 当前行颜色配置如图 1-32 所示。

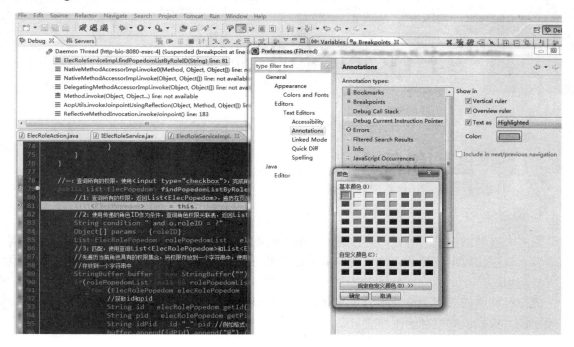

图 1-32　Debug 当前行颜色配置

1.6　小结

本章讲解了 Hadoop 的简介、Hadoop 由来与起源、大事年表，接着介绍了 Hadoop 常用版本优缺点以及生态组件，之后介绍了 Hadoop 中两个重要组件 MapReduce 和 HDFS，最后引导读者安装 Hadoop 的开发环境，包括 Java、Maven、Eclipse。

第 2 章

Hadoop集群搭建

2.1 虚拟机简介

1. 概述

很多人安装虚拟机的时候，经常遇到不能上网的问题，而 VMware 有三种网络模式，对初学者来说也比较眼花缭乱。这里我们基于虚拟机 3 种网络模式，帮大家普及下虚拟机上网的背景知识。

虚拟机网络模式，不论使用的是 VMware、Virtual Box、Virtual PC 等哪种虚拟机软件，一般来说，虚拟机有三种网络模式：

- 桥接
- NAT
- Host-Only

初学者看到虚拟机有三种网络，估计就慌了，笔者也是。哪一种网络是适合自己的虚拟机呢？

2. 桥接

桥接网络是指本地物理网卡和虚拟网卡通过 VMnet0 虚拟交换机进行桥接，物理网卡和虚拟网卡在拓扑图上处于同等地位，那么物理网卡和虚拟网卡就相当于处于同一个网段，虚拟交换机就相当于一台现实网络中的交换机，所以两个网卡的 IP 地址也要设置为同一网段。

所以当我们要在局域网使用虚拟机，对局域网其他 PC 提供服务时，例如提供 FTP 服务、提供 SSH 服务、提供 HTTP 服务，那么就要选择桥接模式。

例如大学宿舍里有一个路由器，宿舍里 4 个人连接这个路由器，路由器的 wan IP 就不理会了，这个 IP 是动态获取的，而 lan IP 默认是 192.168.1.1，子网掩码是 255.255.255.0。而其他 4 个人是自动获取 IP，假设四个人的 IP 是：

- A:192.168.1.100/255.255.255.0
- B:192.168.1.101/255.255.255.0
- C:192.168.1.102/255.255.255.0
- D:192.168.1.103/255.255.255.0

那么虚拟机的 IP 可以设置的 IP 地址是 192.168.1.2~192.168.1.99、192.168.1.104~192.168.1.254（网络地址全 0 和全 1 的除外，再除去 ABCD 四个人的 ip 地址）。

那么虚拟机的 IP 地址可以设置为 192.168.1.98/255.255.255.0，设置了这个 IP 地址，ABCD 4 个人就可以通过 192.168.1.98 访问虚拟机了，如果虚拟机需要上外网，那么还需要配置虚拟机的路由地址，就是 192.168.1.1 了，这样，虚拟机就可以上外网了。但是，上网一般是通过域名去访问外网的，所以还需要为虚拟机配置一个 DNS 服务器，我们可以简单点，把 DNS 服务器地址配置为 Google 的 DNS 服务器:8.8.8.8，到此，虚拟机就可以上网了。

3. NAT

NAT 模式中，就是让虚拟机借助 NAT（网络地址转换）功能，通过宿主机器所在的网络来访问公网。

NAT 模式中，虚拟机的网卡和物理网卡的网络，不在同一个网络，虚拟机的网卡，是在 VMware 提供的一个虚拟网络。

下面我们比较一下 NAT 和桥接。

（1）NAT 模式和桥接模式虚拟机都可以上外网。

（2）由于 NAT 的网络在 VMware 提供的一个虚拟网络里，所以局域网其他主机是无法访问虚拟机的，而宿主机可以访问虚拟机，虚拟机可以访问局域网的所有主机，因为真实的局域网相对于 NAT 的虚拟网络，就是 NAT 的虚拟网络的外网，不懂的人可以查查 NAT 的相关知识。

（3）桥接模式下，多个虚拟机之间可以互相访问；NAT 模式下，多个虚拟机之间也可以相互访问。

如果你建一个虚拟机，只是给自己用，不需要给局域网其他人用，那么可以选择 NAT，毕竟 NAT 模式下的虚拟系统的 TCP/IP 配置信息是由 VMnet8(NAT)虚拟网络的 DHCP 服务器提供的，只要虚拟机的网络配置是 DHCP，那么你不需要进行任何其他的配置，只需要宿主机器能访问互联网，就可以让虚拟机联网了。

例如，你想建多个虚拟机集群，作为测试使用，而宿主机可能是一个笔记本，IP 不固定。这种应用场景，我们需要采用 NAT 模式了，这时我们要注意一个问题，虚拟机之间是需要互访的，默认采用 DHCP，虚拟机的 IP 每次重启，IP 都是不固定的，所以我们需要手工设置虚拟机的 IP 地址。

但是我们对虚拟机网卡所在的虚拟网络的信息还一无所知，例如虚拟机网络的路由地址、子网掩码，所以我们需要先查下 NAT 虚拟网络的信息。

使用 VMware，在 Edit→Virtual Network Editor 中配置好虚拟网络信息，注意 VMnet8，VMnet8 相当于是本机的一个路由，虚拟机设置 NAT 后就是通过这个路由进行上网的，可以查看其网络地址、路由地址、子网掩码。

选择 VMnet8→NAT 设置，可以看到子网 IP 显示为 192.168.233.0，子网掩码是 255.255.255.0，那路由地址呢，其实就是网关 IP 了，都是同一个东西，这里是 192.168.233.2。

接下来就好办了，在对应的虚拟机设置好 IP、子网掩码，路由地址就可以上外网了，至于 DNS 可以设置为 8.8.8.8。

4. Host-Only

在 Host-Only 模式下，虚拟网络是一个全封闭的网络，它唯一能够访问的就是主机。其实 Host-Only 网络和 NAT 网络很相似，不同的地方就是 Host-Only 网络没有 NAT 服务，所以虚拟网络不能连接到 Internet。主机和虚拟机之间的通信是通过 VMware Network Adepter VMnet1 虚拟网卡来实现的。

Host-Only 的宗旨就是建立一个与外界隔绝的内部网络来提高内网的安全性。这个功能或许对普通用户来说没有多大意义，但大型服务商会常常利用这个功能。如果你想为 VMnet1 网段提供路由功能，那就需要使用 RRAS，而不能使用 XP 或 2000 的 ICS，因为 ICS 会把内网的 IP 地址改为 192.168.0.1，但虚拟机是不会给 VMnet1 虚拟网卡分配这个地址的，那么主机和虚拟机之间就不能通信了。

5. 综述

在 VMware 的 3 种网络模式中，NAT 模式是最简单的，基本不需要手动配置 IP 地址等相关参数。至于桥接模式则需要额外的 IP 地址，如果是在内网环境中还很容易，如果是 ADSL 宽带就比较麻烦了，ISP 一般是不会大方地多提供一个公网 IP 的。

2.2 虚拟机配置

1. 前期准备

本节主要讲解如何在 VMware 中配置三台虚拟机互通且联网。

读者可以从【百度云盘】Hadoop 集群目录中直接下载作者配置好的虚拟机（Java 为 1.7 版本，NAT 互通，可以联网）。2.2 节、2.3 节操作已经全部完毕，只要虚拟机加载三台 CentOS 就可以直接跳到 2.4 节开始配置集群。解压后三台虚拟机如图 2-1 所示。

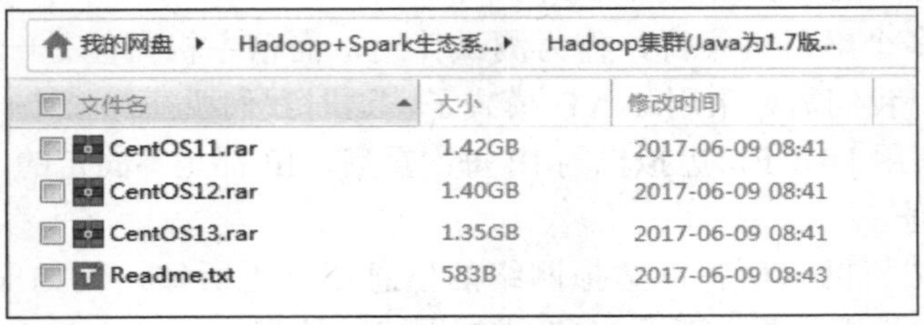

图 2-1 解压后三台虚拟机

- 用户名：root，密码：hadoop。
- 用户名：hadoop，密码：hadoop。

2. 相关软件下载

下载 CentOS6.7，作者已经建立好了一台 Linux 系统的虚拟机，读者只需在【百度云盘】Soft_BigData 目录中下载解压。

- 系统为：CentOS6.7 纯净版。
- 用户名：root，密码：hadoop。
- 用户名：hadoop，密码：hadoop。

其余相关软件下载路径如下：

- Hadoop 和 Spark 程序员软件包

http://blog.csdn.net/silentwolfyh/article/details/50936459

- Java 程序员软件包

http://blog.csdn.net/silentwolfyh/article/details/50936377

3. 下载虚拟机

【百度云盘】Soft_BigData 目录中使用 VMware10.0.4 版本。

4. 加载三个 Linux 虚拟机

将下载的 CentOS 6.7 虚拟机加载出来三个，分别更名为 CentOS01、CentOS02、CentOS03，使用 VMware 加载打开。虚拟机软件加载三台虚拟机整体展示如图 2-2 所示。

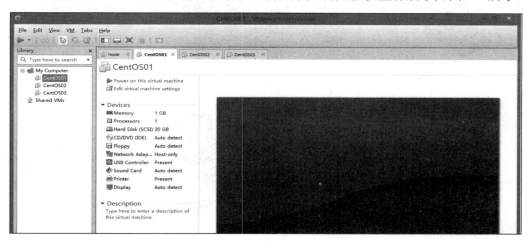

图 2-2　虚拟机软件加载三台虚拟机整体展示

5. 配置多台虚拟机互通且上网

（1）多台配置注意事项

注意：配置过程中会出现一个问题：MAC 地址冲突，所以虚拟机在开启的时候要使用【移动】，切勿使用【复制】。MAC 地址冲突错误如图 2-3 所示。

> ⚠ 适配器 Ethernet0 的 MAC 地址 00:0C:29:3C:BF:E7 属于预留的地址范围，或者正在由系统中的其他虚拟适配器使用。
> 适配器 Ethernet0 可能不具有网络连接。

图 2-3　MAC 地址冲突错误

（2）多台虚拟机配置的架构图

多台虚拟机配置的架构图如图 2-4 所示，这张图是本节的核心，其中包括物理主机的配置、虚拟机软件的配置、三台虚拟主机的配置。

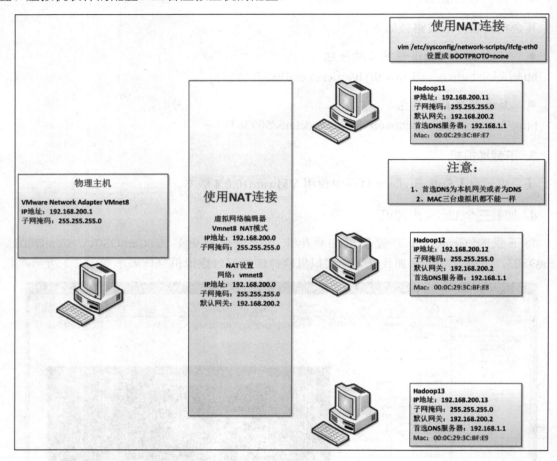

图 2-4　多台虚拟机配置的架构图

（3）物理主机的配置

物理主机的配置如图 2-5 所示。

第 2 章　Hadoop 集群搭建

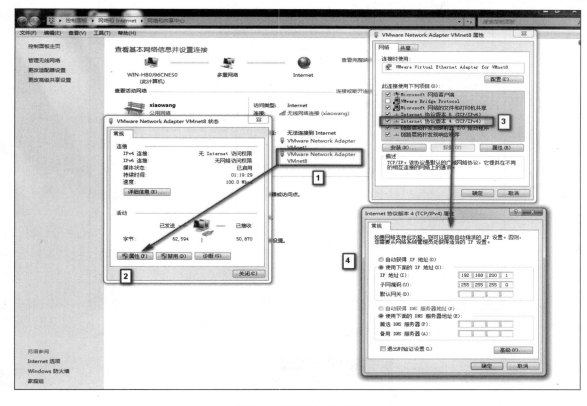

图 2-5　物理主机的配置

（4）虚拟机软件的配置

每台机器设置为 NAT 连接，具体操作如图 2-6~图 2-9 所示。

图 2-6　配置 NAT 连接一

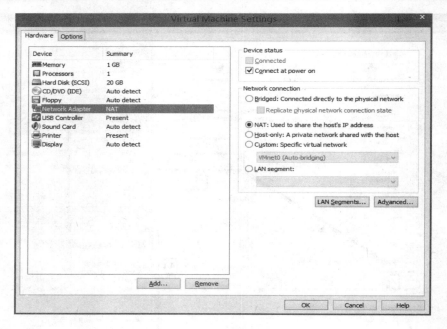

图 2-7　配置 NAT 连接二

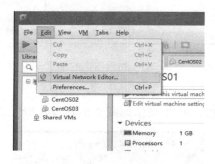

图 2-8　配置 NAT 连接三

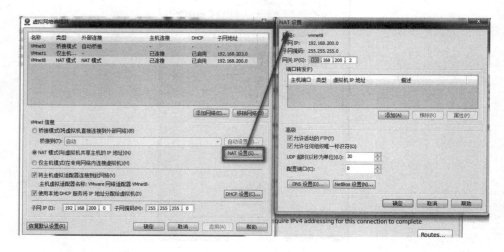

图 2-9　配置 NAT 连接四

（5）三台虚拟主机的配置

开启虚拟机的时候注意：配置过程中会出现下面一个问题：MAC 地址冲突，所以虚拟机在开启的时候要使用移动，切勿使用复制。选择移动，如图 2-10 所示。

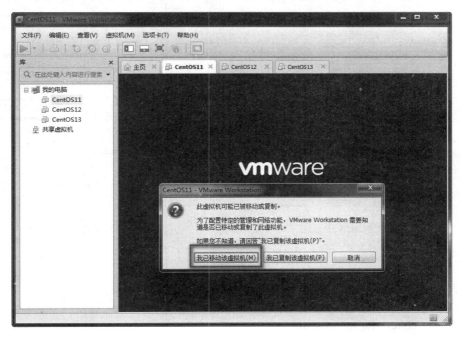

图 2-10　选择移动

如果选择"我已移动该虚拟机"，则这台机器网卡（物理地址）只有一个。

如果选择"我已复制该虚拟机"，则这台机器网卡（物理地址）还需要重新配置，比较麻烦。

进入虚拟机之后配置网络 IP，步骤如图 2-11~2-14 所示。

图 2-11　配置网络 IP 一

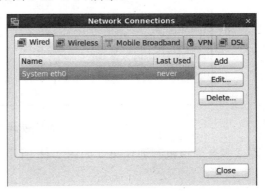

图 2-12　配置网络 IP 二

图 2-13 配置网络 IP 三　　　　　　　图 2-14 配置网络 IP 四

（6）关闭防火墙

- 重启后生效。
  开启：chkconfig iptables on
  关闭：chkconfig iptables off
- 即时生效，重启后失效。
  开启：　service iptables start
  关闭：　service iptables stop

防火墙关闭如图 2-15 所示。

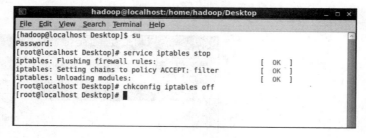

图 2-15 同防火墙关闭

（7）测试多台虚拟机互通结果（如图 2-16 所示）

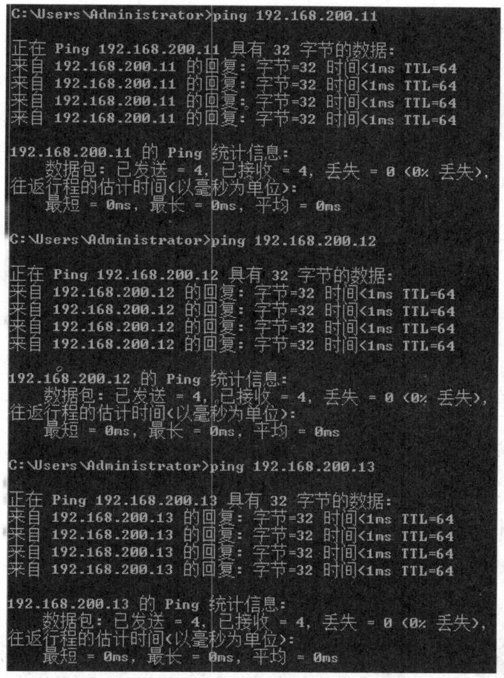

```
C:\Users\Administrator>ping 192.168.200.11

正在 Ping 192.168.200.11 具有 32 字节的数据:
来自 192.168.200.11 的回复: 字节=32 时间<1ms TTL=64
来自 192.168.200.11 的回复: 字节=32 时间<1ms TTL=64
来自 192.168.200.11 的回复: 字节=32 时间<1ms TTL=64
来自 192.168.200.11 的回复: 字节=32 时间<1ms TTL=64

192.168.200.11 的 Ping 统计信息:
    数据包: 已发送 = 4，已接收 = 4，丢失 = 0 (0% 丢失)，
往返行程的估计时间(以毫秒为单位):
    最短 = 0ms，最长 = 0ms，平均 = 0ms

C:\Users\Administrator>ping 192.168.200.12

正在 Ping 192.168.200.12 具有 32 字节的数据:
来自 192.168.200.12 的回复: 字节=32 时间<1ms TTL=64
来自 192.168.200.12 的回复: 字节=32 时间<1ms TTL=64
来自 192.168.200.12 的回复: 字节=32 时间<1ms TTL=64
来自 192.168.200.12 的回复: 字节=32 时间<1ms TTL=64

192.168.200.12 的 Ping 统计信息:
    数据包: 已发送 = 4，已接收 = 4，丢失 = 0 (0% 丢失)，
往返行程的估计时间(以毫秒为单位):
    最短 = 0ms，最长 = 0ms，平均 = 0ms

C:\Users\Administrator>ping 192.168.200.13

正在 Ping 192.168.200.13 具有 32 字节的数据:
来自 192.168.200.13 的回复: 字节=32 时间<1ms TTL=64
来自 192.168.200.13 的回复: 字节=32 时间<1ms TTL=64
来自 192.168.200.13 的回复: 字节=32 时间<1ms TTL=64
来自 192.168.200.13 的回复: 字节=32 时间<1ms TTL=64

192.168.200.13 的 Ping 统计信息:
    数据包: 已发送 = 4，已接收 = 4，丢失 = 0 (0% 丢失)，
往返行程的估计时间(以毫秒为单位):
    最短 = 0ms，最长 = 0ms，平均 = 0ms
```

图 2-16　测试多台虚拟机互通结果

2.3 Linux 系统设置

1. 用户权限事项

配置或者搭建集群过程中，全部用 root 账户登录。

2. 修改主机名方法

将三台机器的主机名分别修改成 hadoop11、hadoop12、hadoop13。修改完之后重启电脑，修改的方法如下：

- 方法一：暂时修改机器名：hostname 。

用 hostname 命令可以临时修改机器名，但机器重新启动之后就会恢复原来的值。

#hostname　　//查看机器名

#hostname -i　　//查看本机器名对应的 ip 地址

- 方法二：永久性修改机器名：修改系统配置文件

修改/etc/sysconfig/network，修改这个系统配置文件，才能有效改变机器名，修改主机名如图 2-17 所示。

图 2-17 修改主机名

3. Windows 和 Linux 域名解析

（1）Windows 的域名解析

修改 Windows 下的域名解析，路径为 C:\Windows\System32\drivers\etc\hosts。

内容为：

```
192.168.200.11 hadoop11
192.168.200.12 hadoop12
192.168.200.13 hadoop13
```

（2）Linux 的域名解析

修改 Linux 下的域名解析如图 2-18 所示，路径为/etc/hosts。

图 2-18 Linux 域名解析设置

4. ssh 免密码登录的配置方法

在客户端生成密钥对：

```
ssh-keygen -t rsa
```

把公钥复制给要登录的目标主机，目标主机上将这个公钥加入到授权列表：

```
Cat id_rsa.pub >>authorized_keys
```

目标主机还要将这个授权列表文件权限修改为：

```
600 chmod 600 authorized_keys
```

在用户目录下面执行 5 步：

```
rm -rf .ssh/
ssh-keygen -t rsa
cat .ssh/id_rsa.pub >> .ssh/authorized_keys
chmod 700 .ssh
chmod 600 .ssh/authorized_keys
```

.ssh/authorized_keys 内容如图 2-19 所示。

```
[root@hadoop11 ~]# cat .ssh/authorized_keys
ssh-rsa AAAAB3NzaC1yc2EAAAABIwAAAQEArzSLUKdz1U3LenyGO1irj5+vpXI9xVRSoLiHi01/BCNnEfnZ2UqrewIPbtvx+
S9Jc9emVS35o9BOKPjv4VFuLkp12kt/gKwXdSaXDSYIBTIX2V0Ge1rFXNPoK+cRcuY6pmmkMmT26goTo1i4B+DEZd8jHTER1E
sBUVaHOXVj1IaYWJEOHRZcNFa5WoMQQJzio8zVuLztAtThnvz4Y/Ck1OmJhSVsW6y0+hJpbzDCosc88Q== root@hadoop11
ssh-rsa AAAAB3NzaC1yc2EAAAABIwAAAQEA7GxVXdeFwHkZdD4bD5KnhUPg+8ctm/IEc4nmSDaF5VbExQI7O9NqunIGwXBK8
YNU1mPWvO0nNyVhzKp/HUdsQGAjeh67ySvHNWt8k0aM2YKwIIH/fZJRT/1CdbbV9+nZFx1DiodQHbzBaKkka94SOqG4vPiV3I
Oqp0f/fdidMx6brBM575rIGcaAtwN9txznM1Zhh438yH4SESqSLqzwc6YQFpYRZuLCAOs+ND6CCSOOJw== root@hadoop12
ssh-rsa AAAAB3NzaC1yc2EAAAABIwAAAQEAxNZv6oHnbe+qWTGnfi2dGrf98MO1IUs1r7Tibe1/Rfj2b6uMCORNNrWMUSTuP
/HxkXs+x9B7SiJKo/IKNihj6yQMW/Lit6+r/6uuroaWKH1sNvGQxk5N3dexKZvONHvQm1OtnRsN26pjHvZg0rEgls7nCWEV38
Cr6F29jqjvQGh5f1i7Eoos51Tbnq3AgKI5GWMtFNojUA159SzSajJsg2oMoI2LqWYXedI8Ua3IGuc23w== root@hadoop13
[root@hadoop11 ~]#
```

图 2-19 authorized_keys 内容

5. Linux 批量关机和重启脚本

为了避免读者每次登录所有机器单独执行某条指令，作者写了一个批量启动脚本。主要方法是远程发送命令，从 A 服务器到达 B 服务器，启动某条 shell 命令。

批量执行的脚本如下：

```
all-halt.sh            //批量关闭所有服务器
all-restart.sh         //批量重启所有服务器
hadoop-halt.sh         //启动一台节点 Hadoop
hadoop-restart.sh      //重启一台节点 Hadoop
```

批量关机脚本 all-halt.sh 如图 2-20 所示。

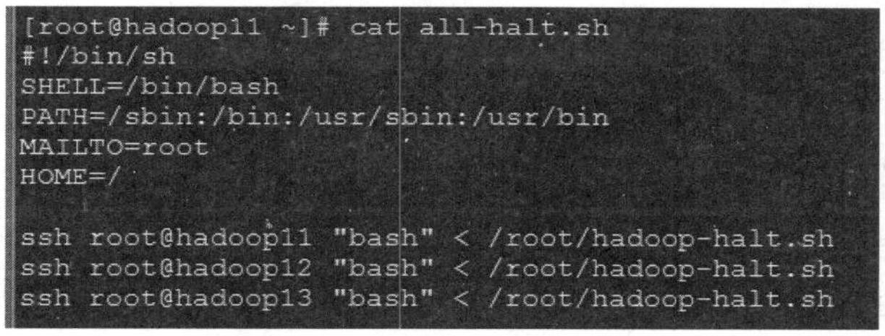

```
[root@hadoop11 ~]# cat all-halt.sh
#!/bin/sh
SHELL=/bin/bash
PATH=/sbin:/bin:/usr/sbin:/usr/bin
MAILTO=root
HOME=/
ssh root@hadoop11 "bash" < /root/hadoop-halt.sh
ssh root@hadoop12 "bash" < /root/hadoop-halt.sh
ssh root@hadoop13 "bash" < /root/hadoop-halt.sh
```

图 2-20 批量关机脚本 all-halt.sh

```
ssh root@hadoop11 "bash" </root/hadoop-halt.sh
ssh root@hadoop12 "bash" </root/hadoop-halt.sh
ssh root@hadoop13 "bash" </root/hadoop-halt.sh
```

6. Linux 虚拟机安装 Java

每台机器安装 JDK，下载 Java 版本 jdk-8u65-linux-x64.tar.gz。

（1）卸载系统自带 Java

```
[root@hadoop11 app]#  rpm -qa|grep java
[root@hadoop11 app]#  yum -y remove java-1.7.0-openjdk-1.7.0.79-
2.5.5.4.el6.x86_64
```

（2）安装 Java

配置环境变量：

```
vi /etc/profile
export JAVA_HOME=/usr/app/jdk1.8.0_65
export PATH=$PATH:$JAVA_HOME/bin
```

立即生效：

```
source /etc/profile
```

7. 所有节点配置 NTP 服务

集群中所有主机必须保持时间同步，如果时间相差较大会引起各种问题。具体思路如下：

Master 节点作为 NTP 服务器与外界的对时中心同步时间，随后对所有 DataNode 节点提供时间同步服务。所有 DataNode 节点以 Master 节点为基础同步时间。

所有节点安装相关组件：

```
yum install ntp
```

完成后，配置开机启动：

```
chkconfig ntpd on
```

检查是否设置成功：

```
chkconfig --list ntpd
```

其中从第 2~5 个运行级系统服务转态为 on 状态就代表成功。

（1）主节点配置

在配置之前，先使用 ntpdate 手动同步一下时间，免得本机与对时中心时间差距太大，使得 ntpd 不能正常同步。这里选用 127.127.1.0 作为对时中心：

```
ntpdate -u 127.127.1.0
```

NTP 服务只有一个配置文件（NTP 配置文档只有一个 /etc/ntp.conf，看看作者的 ntp.conf），配置好了就可以了。这里只给出有用的配置，不需要的配置都用#注释掉，这里注释掉的内容就不再给出。

Master 的 NTP 配置如下：

```
driftfile /var/lib/ntp/drift
restrict 127.0.0.1
restrict -6 ::1
restrict default nomodify notrap
server 127.127.1.0 prefer
includefile /etc/ntp/crypto/pw
keys /etc/ntp/keys
```

（2）配置 NTP 客户端（所有 DataNode 节点）

DataNode 的 NTP 配置如下：

```
driftfile /var/lib/ntp/drift
restrict 127.0.0.1
```

```
restrict -6 ::1
restrict default kod nomodify notrap nopeer noquery
restrict -6 default kod nomodify notrap nopeer noquery

#这里是主节点的主机名或者ip
server hadoop11
include file /etc/ntp/crypto/pw
keys /etc/ntp/keys
```

启动 NTP 服务：

```
Service ntpd start
```

登录 Master 机器，执行命令：

```
ntpdate -u 127.127.1.0
```

登录所有 Slave 机器，同步主节点执行命令：

```
ntpdate -u hadoop11
```

检查是否成功，查看命令：

```
watch ntpq -p
```

NTP 测试如图 2-21~图 2-23 所示，本例中 hadoop4 是主节点 Master，hadoop1 和 hadoop3 是从节点 Slave，图 2-21 是 hadoop4 查看界面，图 2-22 是 hadoop1 查看界面，图 2-23 是 hadoop3 查看界面。

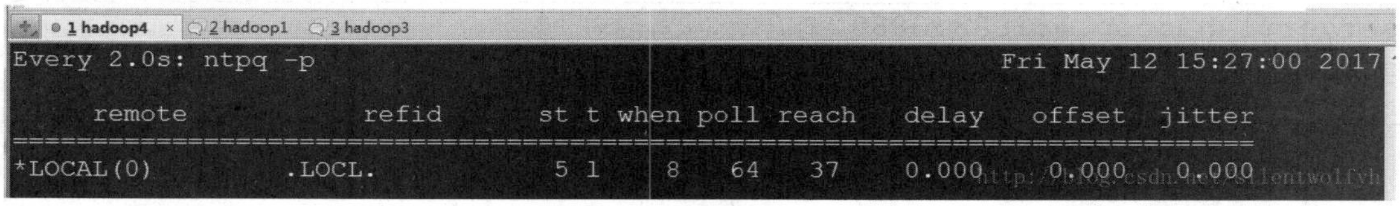

图 2-21　NTP 测试一

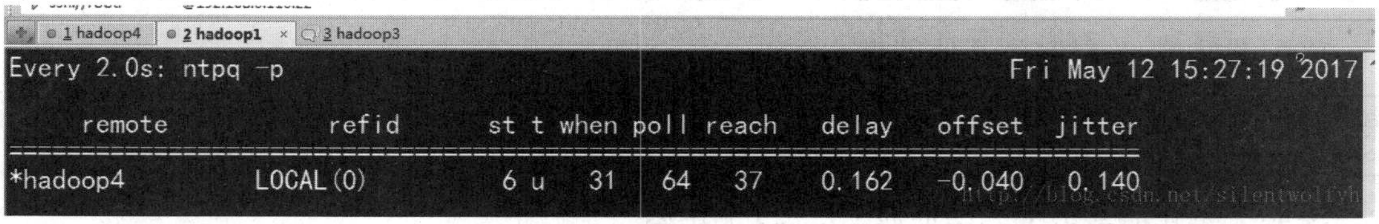

图 2-22　NTP 测试二

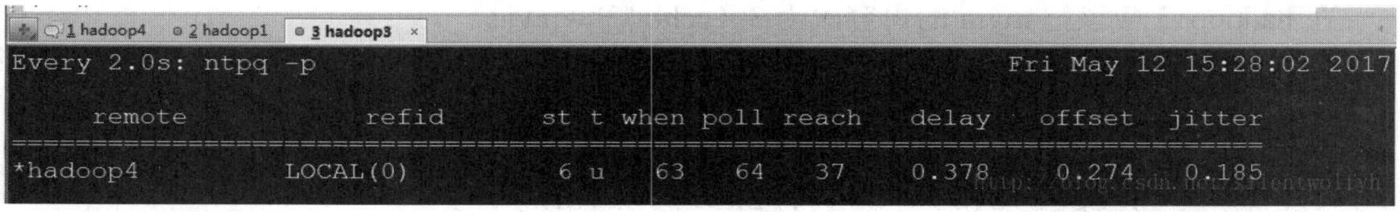

图 2-23　NTP 测试三

2.4 Apache 版本 Hadoop 集群搭建

Apache 版本的所有需要安装的软件全部放在/usr/app 目录下。

1. Zookeeper 安装

以下是 ZooKeeper 的安装步骤：

（1）上传 zk 安装包。

（2）解压。

```
tar -zxvf  zookeeper-3.4.6.tar.gz  /usr/app/
```

（3）配置（先在一台节点上配置）。

添加一个 zoo.cfg 配置文件：

```
cd zookeeper-3.4.6/conf/
cp -r zoo sample.cfg zoo.cfg
```

修改配置文件（zoo.cfg）：

```
mkdir /usr/app/zookeeper-3.4.6/data
dataDir=/usr/app/zookeeper-3.4.6/data(the directory where the
snapshot is stored.)
```

在最后一行添加：

```
server.1=hadoop11:2888:3888
server.2=hadoop12:2888:3888
server.3=hadoop13:2888:3888
```

在（dataDir=/usr/app/zookeeper-3.4.5/data）创建一个 myid 文件，里面内容是 server.N 中的 N（server.2 里面内容为 2）。

```
echo "1" >myid
```

将配置好的 zk 复制到其他节点。

```
scp  -r  /usr/app/zookeeper-3.4.6/   root@hadoop12:/usr/app
scp  -r  /usr/app/zookeeper-3.4.6/   root@hadoop13:/usr/app
```

注意：在其他节点上一定要修改 myid 的内容。

在 hadoop12 应该把 myid 的内容改为 2（echo "2" >myid）。

在 hadoop13 应该把 myid 的内容改为 3（echo "3" >myid）。

（4）启动 ZooKeeper 集群。

三台机器上面分别启动 ZooKeeper，启动和关闭命令如下：

```
/usr/app/zookeeper-3.4.6/bin/./zkServer.sh start
/usr/app/zookeeper-3.4.6/bin/./zkServer.sh stop
```

每台机器分别启动 zk，如图 2-24 所示。

```
./zkServer.sh start
```

图 2-24　每台机器分别启动 zk

（5）查看启动状态。

```
/usr/app/zookeeper-3.4.6/bin/./zkServer.sh status
```

每台 ZooKeeper 会选出 leader 或者 follower。ZooKeeper 的 follwer 状态如图 2-25 所示，ZooKeeper 的 leader 状态如图 2-26 所示。

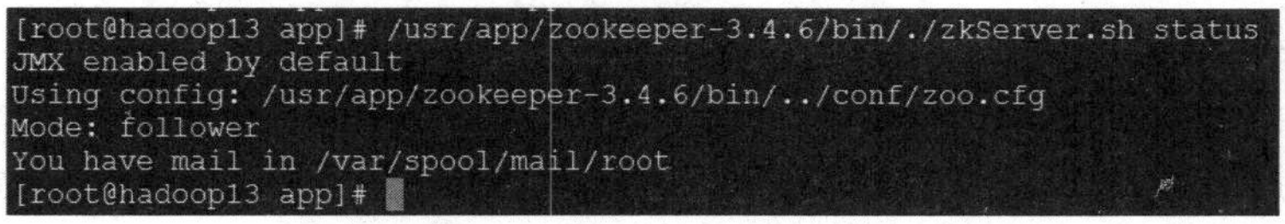

图 2-25　ZooKeeper 的 follwer 状态

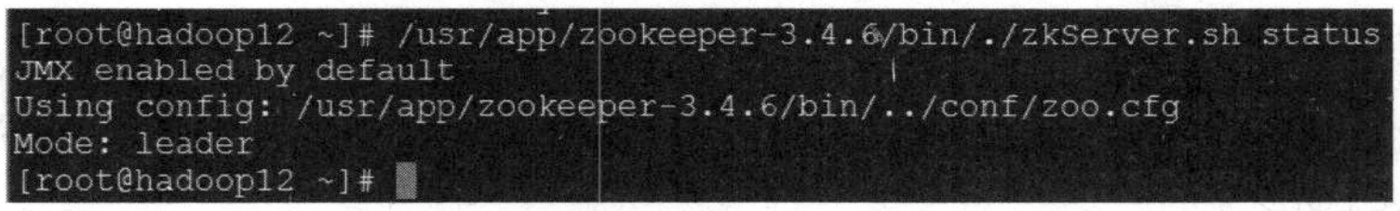

图 2-26　ZooKeeper 的 leader 状态

2. 安装 Hadoop

下载 hadoop-2.6.0.tar.gz 压缩包。

（1）解压 hadoop-2.6.0.tar.gz 压缩包。

```
tar -xzvf hadoop-2.6.0.tar.gz
```

（2）配置环境变量：vi /etc/profile。

```
export JAVA_HOME=/usr/app/jdk1.7
export HADOOP_HOME=/usr/app/hadoop-2.6.0
export PATH=$PATH:$JAVA_HOME/bin:$HADOOP_HOME/bin
```

（3）配置 hadoop-env.sh 文件，修改 JAVA_HOME。

/usr/app/hadoop-2.6.0/etc/hadoop

```
# The java implementation to use.
export JAVA_HOME=/usr/app/jdk1.7
```

（4）配置 slaves 文件，增加 slave 节点。

/usr/app/hadoop-2.6.0/etc/hadoop

```
hadoop11
hadoop12
hadoop13
```

(5)配置 core-site.xml 文件

core-site.xml

```
<configuration>
<property>
<name>fs.defaultFS</name>
<value>hdfs://ns1</value>
</property>
【这里的值指的是默认的 HDFS 路径。当有多个 HDFS 集群同时工作时,集群名称在这里指
定!该值来自于 hdfs-site.xml 中的配置】

<property>
<name>hadoop.tmp.dir</name>
<value>/usr/app/hadoop-2.6.0/tmp</value>
</property>
【这里的路径默认是 NameNode、DataNode、JournalNode 等存放数据的公共目录。用户
也可以自己单独指定这三类节点的目录。】

<property>
<name>ha.zookeeper.quorum</name>
<value>hadoop11:2181,hadoop12:2181,hadoop13:2181</value>
</property>
【这里是 ZooKeeper 集群的地址和端口。注意,数量一定是奇数,且不少于三个节点】
</configuration>
```

(6)配置 hdfs-site.xml 文件

hdfs-site.xml

```
<configuration>
<property>
<name>dfs.replication</name>
<value>3</value>
</property>
【指定 DataNode 存储 block 的副本数量。默认值是3个,我们现在有4个 DataNode,该值
不大于4即可。】

<property>
<name>dfs.nameservices</name>
<value>ns1</value>
</property>
【使用 federation 时,HDFS 集群别名。名字可以随便起,多个集群时相互不重复即可】

<property>
<name>dfs.ha.namenodes.ns1</name>
<value>nn1,nn2</value>
</property>
【指定该集群的 namenode 的机器】
```

```xml
<property>
<name>dfs.namenode.rpc-address.ns1.nn1</name>
<value>hadoop11:9000</value>
</property>
```
【指定 hadoop100 的 RPC 地址】

```xml
<property>
<name>dfs.namenode.http-address.ns1.nn1</name>
<value>hadoop11:50070</value>
</property>
```
【指定 hadoop100 的 http 地址】

```xml
<property>
<name>dfs.namenode.rpc-address.ns1.nn2</name>
<value>hadoop12:9000</value>
</property>
```
【指定 hadoop101 的 RPC 地址】

```xml
<property>
<name>dfs.namenode.http-address.ns1.nn2</name>
<value>hadoop12:50070</value>
</property>
```
【指定 hadoop101 的 http 地址】

```xml
<property>
<name>dfs.namenode.shared.edits.dir</name>
<value>qjournal://hadoop11:8485;hadoop12:8485;hadoop13:8485/ns1</value>
</property>
```
【指定该集群的两个 NameNode 共享 edits 文件目录时，使用的 JournalNode 集群信息】

```xml
<property>
<name>dfs.journalnode.edits.dir</name>
<value>/usr/app/hadoop-2.6.0/journaldata</value>
</property>
```
【指定该集群是否启动自动故障恢复，即当 NameNode 出故障时，是否自动切换到另一台 NameNode】

```xml
<property>
<name>dfs.ha.automatic-failover.enabled</name>
<value>true</value>
</property>
```
【指定该集群出故障时，哪个实现类负责执行故障切换】

```xml
<property>
<name>dfs.client.failover.proxy.provider.ns1</name>
<value>org.apache.hadoop.hdfs.server.namenode.ha.ConfiguredFailoverProxyProvider</value>
</property>
```

【client 的 failover 代理配置】

```
<property>
<name>dfs.ha.fencing.methods</name>
<value>sshfence</value>
</property>
```
【一旦需要 NameNode 切换，使用 ssh 方式进行操作】

```
<property>
<name>dfs.ha.fencing.ssh.private-key-files</name>
<value>/root/.ssh/id_rsa</value>
</property>
```
【如果使用 ssh 进行故障切换，使用 ssh 通信时用的密钥存储的位置】

```
<property>
<name>dfs.ha.fencing.ssh.connect-timeout</name>
<value>30000</value>
</property>
```
【connect-timeout 连接超时】
```
</configuration>
```

（7）配置 mapred-site.xml 文件

mapred-site.xml

```
<configuration>
<property>
<name>mapreduce.framework.name</name>
<value>yarn</value>
</property>
```
【指定运行 mapreduce 的环境是 yarn，与 hadoop1 截然不同的地方】
```
</configuration>
```

（8）配置 yarn-site.xml 文件

yarn-site.xml

```
<configuration>
<property>
<name>yarn.resourcemanager.ha.enabled</name>
<value>true</value>
</property>
```
【启动 HA 高可用性】

```
<property>
<name>yarn.resourcemanager.cluster-id</name>
<value>yrc</value>
</property>
```
【指定 resourcemanager 的名字】

```
<property>
<name>yarn.resourcemanager.ha.rm-ids</name>
<value>rm1,rm2</value>
```

```xml
</property>
【使用了2个resourcemanager,分别指定Resourcemanager的地址】

<property>
<name>yarn.resourcemanager.hostname.rm1</name>
<value>hadoop11</value>
</property>
【自定ResourceManager1的地址】

<property>
<name>yarn.resourcemanager.hostname.rm2</name>
<value>hadoop12</value>
</property>
【自定ResourceManager2的地址】

<property>
<name>yarn.resourcemanager.zk-address</name>
<value>hadoop11:2181,hadoop12:2181,hadoop13:2181</value>
</property>
【制定ZooKeeper机器】

<property>
<name>yarn.nodemanager.aux-services</name>
<value>mapreduce_shuffle</value>
</property>
【默认】
</configuration>
```

(9) 将配置好的 Hadoop 文件复制到另一台 slave 机器上。

```
[root@hadoop11 app]# scp -r hadoop-2.6.0  root@hadoop13:/usr/app
```

(10) 启动 journalnode（分别在 hadoop11、hadoop12、hadoop13 上执行）。

```
cd /usr/app/hadoop-2.6
   sbin/hadoop-daemon.sh start  journalnode
```

运行 jps 命令检验，hadoop11、hadoop12、hadoop13 上多了 JournalNode 进程。

(11) 格式化 HDFS。

在 hadoop11 的/usr/app/hadoop-2.6.0/bin 执行命令：

```
hdfs  namenode  -format
```

格式化后会根据 core-site.xml 中的 hadoop.tmp.dir 配置生成文件，这里作者配置的是 /usr/app/hadoop-2.6/tmp，然后将/usr/app/hadoop-2.6/tmp 复制到 hadoop12 的/hadoop-2.4.1/下。

```
scp  -r  /usr/app/hadoop-2.6/tmp/   hadoop12:/usr/app/hadoop-2.6/
```

也可以这样，在 hadoop12 上执行命令：

```
hdfs  namenode  -bootstrap  Standby
```

HDFS 格式化成功，如图 2-27 所示。

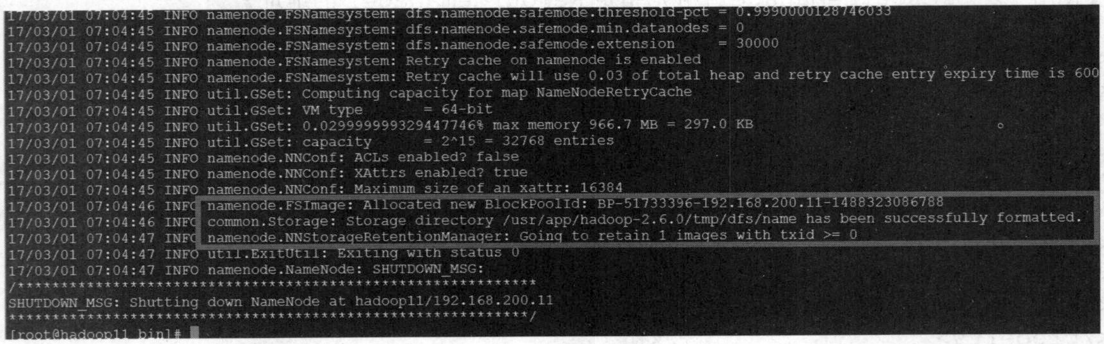

图 2-27　HDFS 格式化成功

（12）格式化 ZKFC（在 hadoop11 上执行即可）

ZooKeeper 格式化命令：hdfs zkfc -formatZK 。ZooKeeper 格式化成功，如图 2-28 所示。

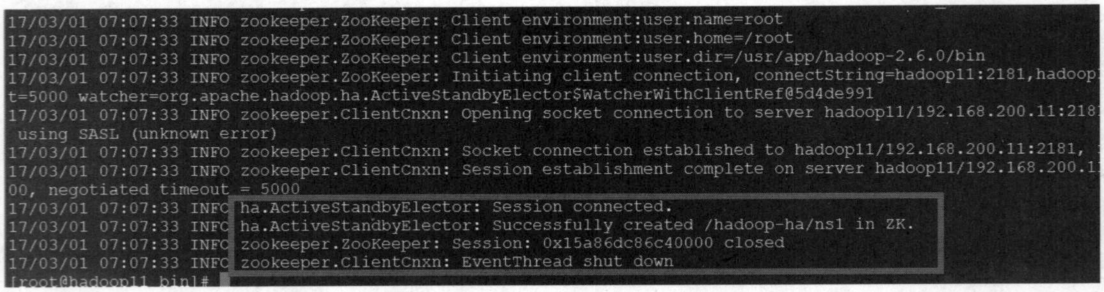

图 2-28　ZooKeeper 格式化成功

（13）启动 HDFS（在 hadoop11 上执行）。

启动 HDFS 命令：sbin/start-dfs.sh 。 HDFS 目录界面如图 2-29 所示

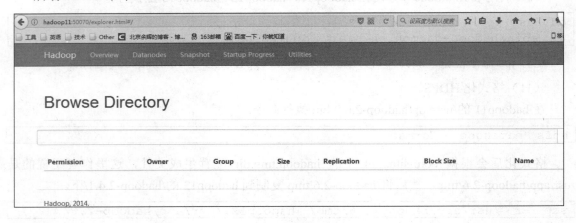

图 2-29　HDFS 目录界面

（14）启动 YARN。注意：是在 hadoop11 上执行 start-yarn.sh，把 namenode 和 resourcemanager 分开是因为性能问题，因为他们都要占用大量资源，所以就要分别在不同的

机器上启动。

启动 YARN 的命令：sbin/start-yarn.sh。YARN 界面如图 2-30 所示。

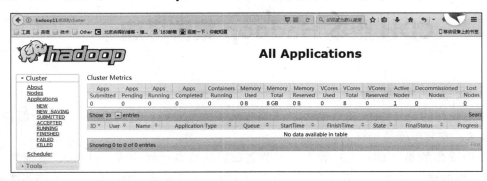

图 2-30　YARN 界面

（15）验证 HDFS HA

查看 HDFS：http://hadoop11:50070/。

查看 RM：http://hadoop11:8088/。

首先向 HDFS 上传一个文件。

```
hadoop fs -put /etc/profile /
hadoop fs -ls /
```

然后再 kill 掉 active 的 NameNode。

```
kill -9 <pid of NN>
```

通过浏览器访问：http://192.168.200.12:50070。

```
NameNode hadoop12:9000' (active)
```

这个时候 hadoop12 上的 NameNode 变成了 active。

执行命令：

```
hadoop fs -ls /
-rw-r--r--   3 root supergroup       1926 2017-02-06 15:36 /profile
```

刚才上传的文件依然存在！！！

手动启动那个挂掉的 NameNode。

```
sbin/hadoop-daemon.sh start namenode
```

通过浏览器访问：http://192.168.200.11:50070。

```
NameNode hadoop12:9000' (standby)
```

（16）验证 YARN：

运行一下 hadoop 提供的 demo 中的 WordCount 程序：

```
hadoop jar share/hadoop/mapreduce/hadoop-mapreduce-examples-
2.4.1.jar wordcount /profile /out
```

OK，大功告成！！！

2.5 CDH 版本 Hadoop 集群搭建

2.5.1 安装前期准备

本节中操作的是 CDH5.10.0 版本。下文安装步骤是曾经安装成功的，之后升级到 CDH 5.10.0 版本，安装步骤是一样的。

本节中安装的是 CDH5.1.3 版本，安装之前默认读者已经完成了 Linux 系统的设置。CDH 对硬件的要求是每台服务器内存不低于 10GB，所以安装 CDH 版本的主机不能是虚拟机。本节有任何安装问题可以参考作者博文：

- CDH 安装和维护

http://blog.csdn.net/silentwolfyh/article/details/54893826

- CDH 问题及维护汇总

http://blog.csdn.net/silentwolfyh/article/details/54893826

1. 下载 CDH 的 Hadoop 版本软件

下载地址：

http://archive.cloudera.com/cdh5/parcels/5.10.0/

需要下载操作系统对应的版本，如图 2-31 所示。也可以从【百度云盘】中下载。

- CDH-5.10.0-1.cdh5.10.0.p0.41-el7.parcel
- CDH-5.10.0-1.cdh5.10.0.p0.41-el7.parcel.sha1
- manifest.json

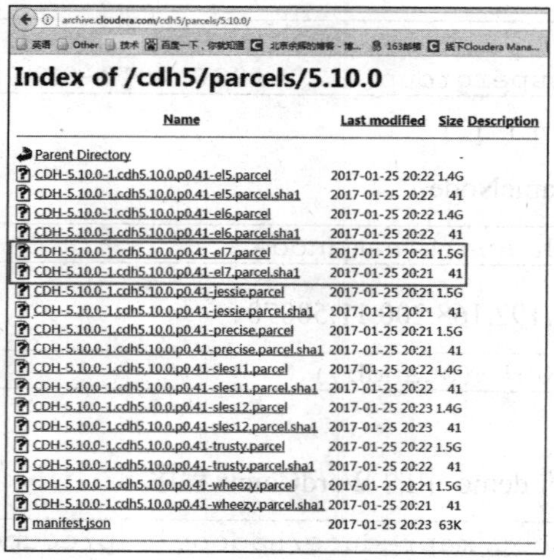

图 2-31 CDH 版本

2. 安装必备软件

```
yum -y install psmisc
yum -y install libxslt
yum -y install screen
yum -y install telnet
```

3. 安装 MySQL

登录 Master 机器，若可以联网，执行命令：

```
yum install mysql-server
```

添加为自启动：

```
chkconfig mysqld on
```

启动 MySQL：

```
service mysqld start
mysqladmin -u root password '123456'
```

进入 mySQL：

```
mysql -uroot -p123456
```

执行下面四句 SQL 语句：

```
create database hive DEFAULT CHARSET utf8 COLLATE utf8_general_ci;
create database ooz DEFAULT CHARSET utf8 COLLATE utf8_general_ci;
create database amon DEFAULT CHARSET utf8 COLLATE utf8_general_ci;
create database hue default charset utf8 collate utf8_general_ci;
```

执行授权语句：

```
grant all privileges on *.* to 'root'@'%' identified by '123456' with
grant option;

flush privileges;
```

2.5.2 Cloudera Manager 安装

1. 创建 cloudera-scm 用户

登录集群所有机器，执行命令，创建 Cloudera SCM 用户：

```
useradd --system --home=/opt/cm-5.1.3/run/cloudera-scm-server/ --no-
create-home --shell=/bin/false --comment "Cloudera SCM User"
cloudera-scm
```

2. 安装 Cloudera Manager

登录 Master 机器，将 cloudera-manager-el6-cm5.1.3_x86_64.tar.gz 文件解压至/opt 下，产

生 2 个文件夹 cloudera 和 cm-5.1.3。

3. 添加 MySQL Connector 包

登录 Master 机器，将 mysql-connector-java-5.1.34-bin.jar 文件复制至目录 /opt/cm-5.1.3/share/ cmf/lib/。

4. 初始化 CM5 的数据库

登录 Master 机器，执行命令：

```
/opt/cm-5.1.3/share/cmf/schema/scm_prepare_database.sh mysql cm -hlocalhost -uroot -p123456 --scm-host localhost scm scm
```

5. 修改 Agent 配置

登录 Master 机器，执行命令：

```
vi /opt/cm-5.1.3/etc/cloudera-scm-agent/config.ini
```

将 server_host 的值改成 Master 机器的名称同步 Agent 到其他节点。

登录 Master 机器，执行命令：

```
scp -r /opt/cm-5.1.3 root@slave[1-xx]:/opt/
```

6. 准备 Parcels

登录 Master 机器，将 CDH-5.1.3-1.cdh5.1.3.p0.12-el6.parcel、CDH-5.1.3-1.cdh5.1.3.p0.12-el6.parcel.sha1 和 Manifest.json 三个文件复制至/opt/cloudera/parcel-repo/文件夹下。

并将 CDH-5.1.3-1.cdh5.1.3.p0.12-el6.parcel.parcel.sha1 重命名为 CDH-5.1.3-1.cdh5.1.3.p0.12-el6.parcel.parcel.sha。

7. 启动 Cloudera Manager

启动 Cloudera Manager 后等待 1~3 分钟，然后访问 http://主节点 IP:7180，若可以访问则表示安装成功。

登录 Master 机器，执行命令：

```
/opt/cm-5.1.3/etc/init.d/cloudera-scm-server start
```

登录集群所有机器，执行命令：

```
/opt/cm-5.1.3/etc/init.d/cloudera-scm-agent start
```

2.5.3 CDH 安装

1. 第一步登录

如图 2-32 所示为登录界面，用户名、密码皆设为 admin。

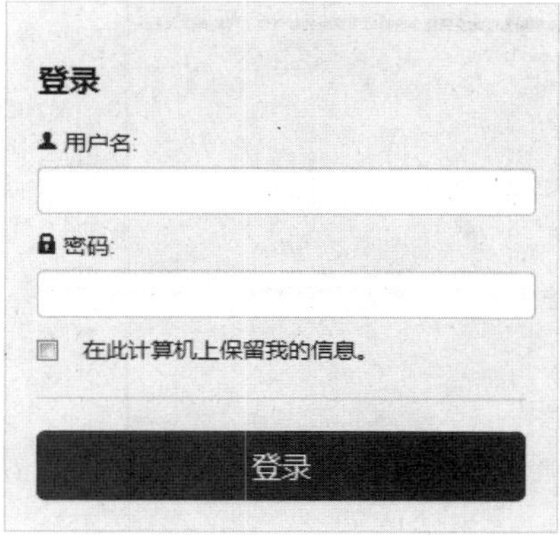

图 2-32

2. 第二步选择部署版本

如图 2-33、图 2-34 所示为选择部署版本。

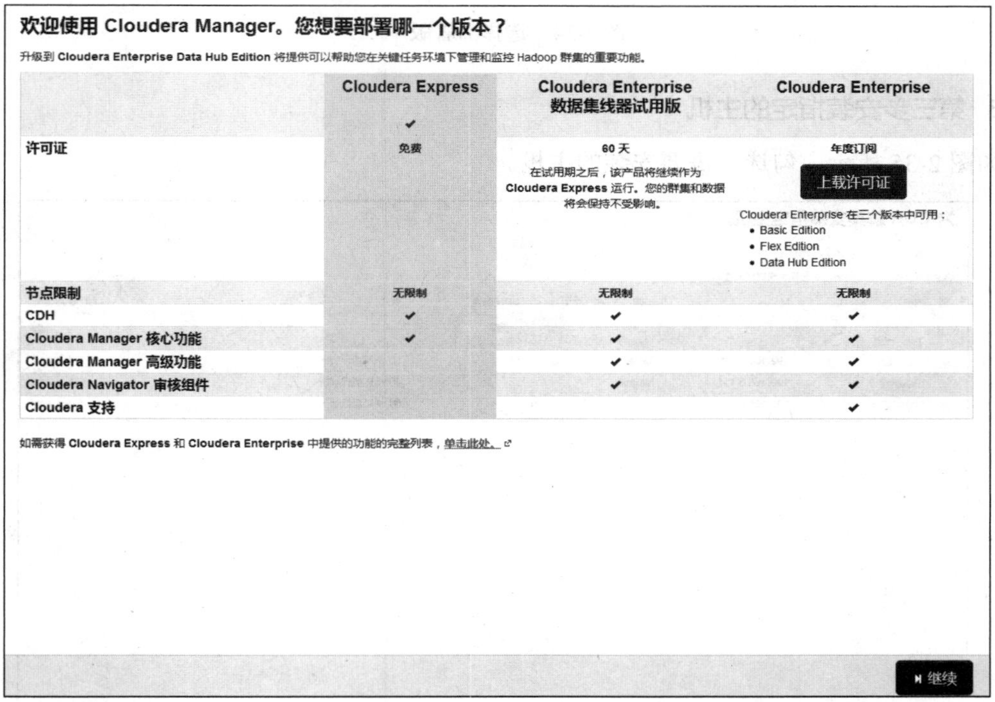

图 2-33　选择部署版本一

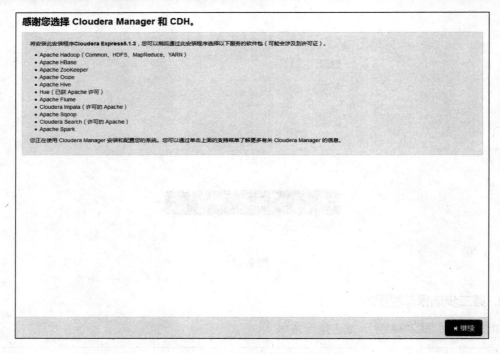

图 2-34　选择部署版本二

3. 第三步安装指定的主机

如图 2-35 所示，勾选上需要安装的主机。

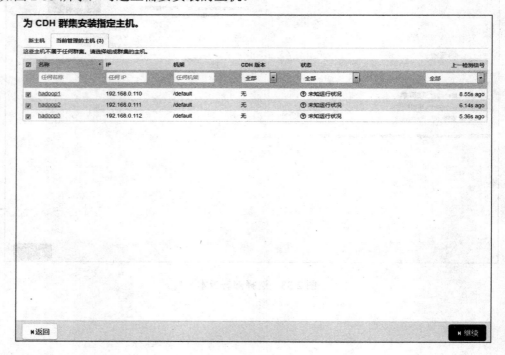

图 2-35　安装指定的主机

4. 第四步集群版本选择

如图 2-36 所示,可以看到所下载并安装的 CDH 版本表示前面部署成功,否则需要在线下载。

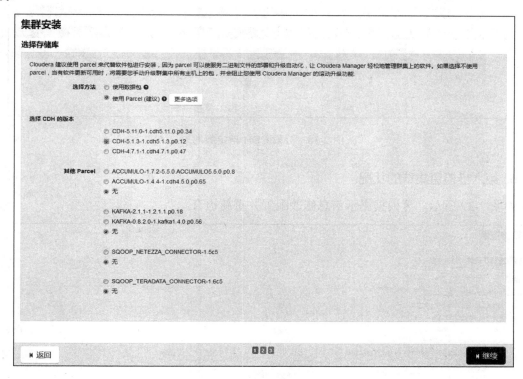

图 2-36　集群版本选择

5. 第五步分发 CDH 指定版本

前面配置无误则无须下载,否则表示前面配置有误。分发可能需要一段时间,视集群大小而定(此时集群一定要保证 Master 可以免密码登录到其他机器上),如图 2-37、图 2-38 所示。

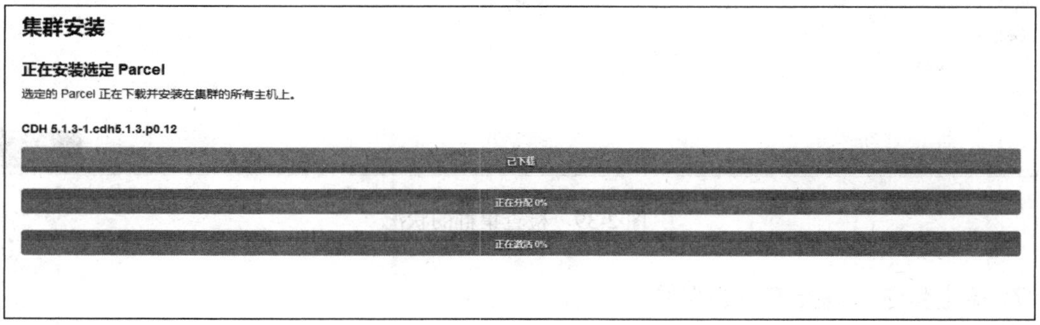

图 2-37　分发 CDH 指定版本一

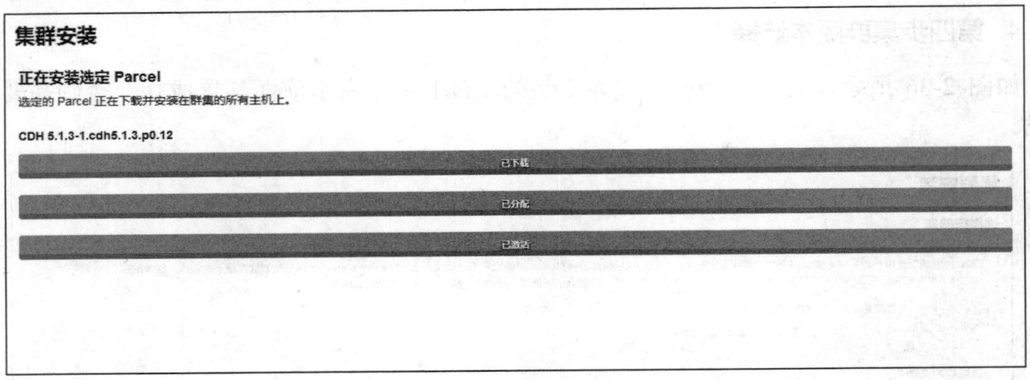

图 2-38　分发 CDH 指定版本二

6. 第六步检查集群的状况

如图 2-39 所示，可根据提示信息修改配置并重新检查。

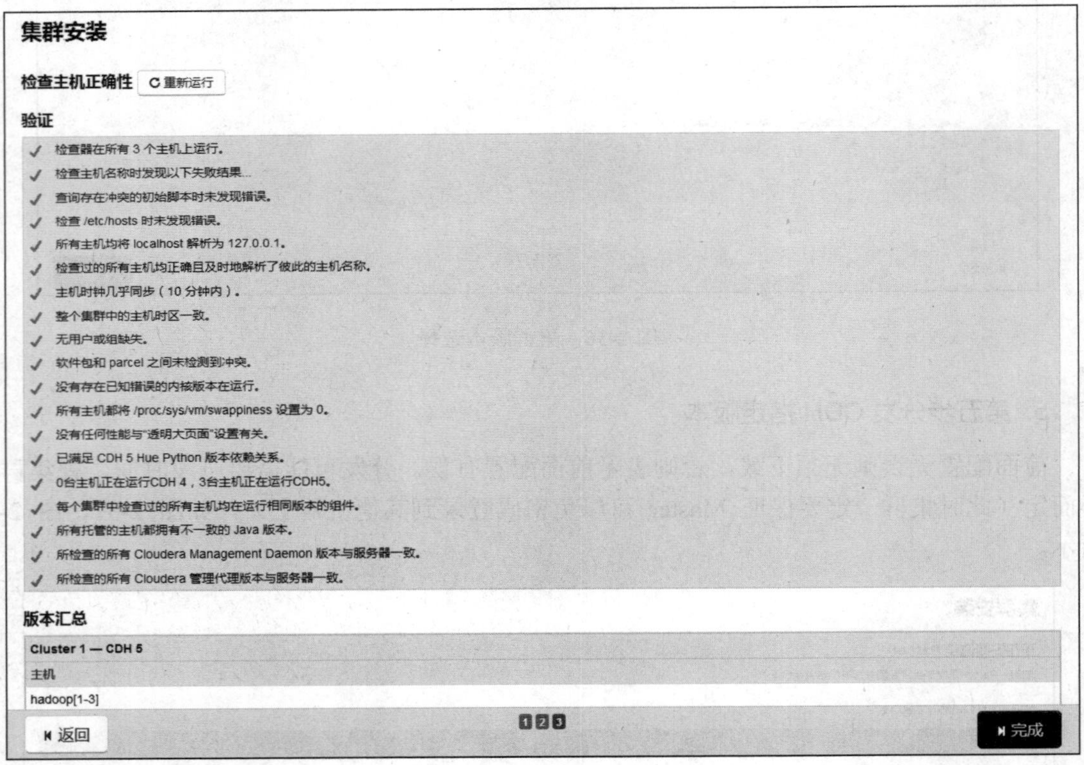

图 2-39　检查集群的状况

7. 第七步选择需要安装的组件

选择需要安装的组件，可自定义安装，如图 2-40 所示。

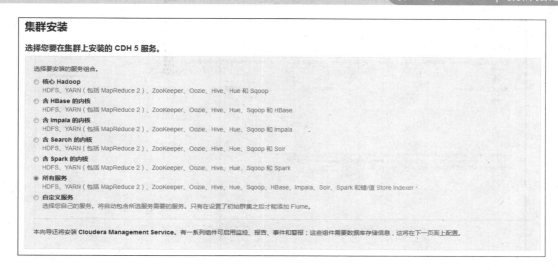

图 2-40　选择需要安装的组件

8. 第八步给集群各个节点分配角色

目前所知道需要注意的地方是 HBase Thrift Server 不要为空，否则 Hue 无法访问 HBase。主要看 Hive 的角色，如图 2-41 所示。

图 2-41　给集群各个节点分配角色

9. 第九步设置 MySQL 连接

如图 2-42 所示。注意：这里很容易报错，都是表权限、名称、密码的错误。

图 2-42 设置 MySQL 连接

10. 第十步完成集群设置

后面的一路"继续"下去即可。安装所需时间较长，请耐心等待。设置过程如图 2-43~图 2-46 所示。

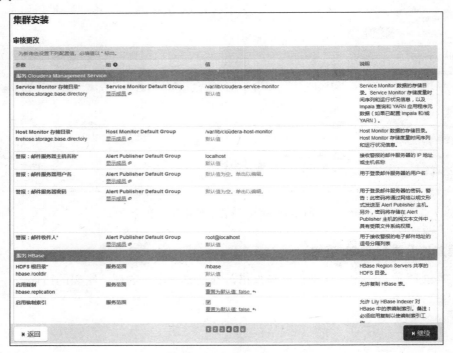

图 2-43 完成集群设置一

图 2-44　完成集群设置二

图 2-45　完成集群设置三

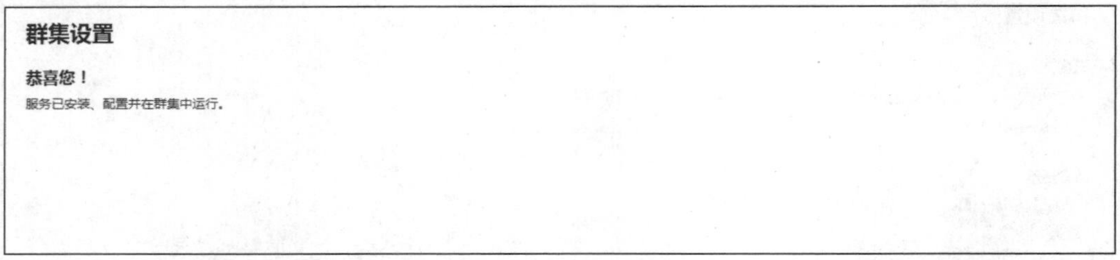

图 2-46　完成集群设置四

11. CDH5.10.0 主界面

CDH5.10.0 主界面如图 2-47 所示。

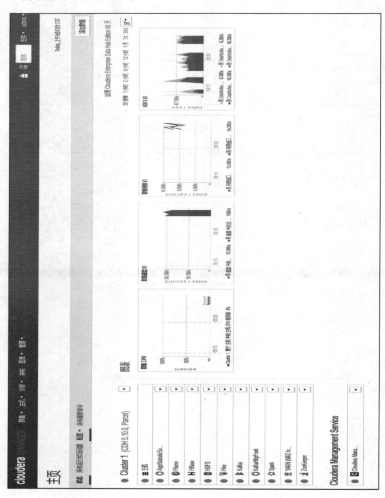

图 2-47　CDH5.10.0 主界面

2.6 小结

本章主要讲解如何配置 Apache 版本和 CDH 版本的 Hadoop 集群。Apache 版本的 Hadoop 集群对于硬件要求低，可以在单台内存 16GB 的机器中建立 3 台虚拟机搭建伪分布式 Hadoop 集群，因此本章中也讲解了虚拟机的使用。CDH 版本的 Hadoop 集群对硬件要求高，只能使用实体机搭建，因此作者使用了 4 台 16GB 的实体机搭建 CDH。为了让读者熟悉这两个版本的 Hadoop 平台，本书后面从第 3~7 章是在 Apache 版本下的 Hadoop 平台进行操作，从第 8~12 章是在 CDH 版本下的 Hadoop 进行操作。

第 3 章

Hadoop 基础与原理

3.1 MapReduce 原理介绍

3.1.1 MapReduce 的框架介绍

MapReduce 模型主要包含 Mapper 类和 Reducer 类两个抽象类。Mapper 类主要负责对数据的分析处理，最终转化为 key-value 数据对；Reducer 类主要获取 key-value 数据对，然后处理统计，得到结果。MapReduce 实现了存储的均衡，但没有实现计算的均衡。MapReduce 框架如图 3-1 所示。

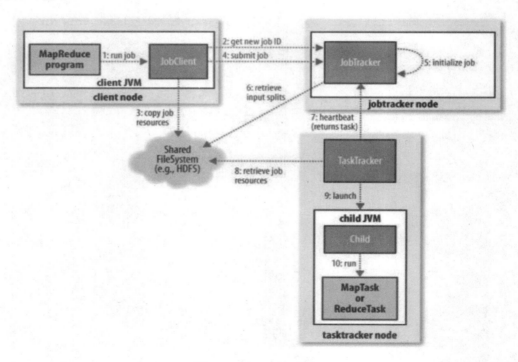

图 3-1 MapReduce 框架

MapReduce 主要包括 JobClient、JobTracker、TaskTracker、HDFS 4 个独立的部分。

1. JobClient

配置参数 Configuration，并打包成 jar 文件存储在 HDFS 上，将文件路径提交给 JobTracker 的 Master 服务，然后由 Master 创建每个 task 将它们分发到各个 TaskTracker 服务中去执行。

2. JobTracker

这是一个 Master 服务，程序启动后，JobTracker 负责资源监控和作业调度。JobTracker 监控所有的 TaskTracker 和 job 的健康状况，一旦发生失败，即将之转移到其他节点上，同时 JobTracker 会跟踪任务的执行进度、资源使用量等信息，并将这些信息告诉任务调度器，而调度器会在资源出现空闲时，选择合适的任务使用这些资源。在 Hadoop 中，任务调度器是一个可插拔的模块，用户可以根据自己的需要设计相应的调度器。

3. TaskTracker

TaskTracker 是运行在多个节点上的 slaver 服务。TaskTracker 主动与 JobTracker 通信，接受作业，并负责直接执行每个任务。TaskTracker 会周期性地通过 Heartbeat 将本节点上资源的使用情况和任务的运行进度汇报给 JobTracker，同时接收 JobTracker 发送过来的命令，并执行相应的操作（如启动新任务、杀死任务等）。TaskTracker 使用"slot"等量划分本节点上的资源量。"slot"代表计算资源（CPU、内存等）。一个 Task 获取到一个 slot 后才有机会运行，而 Hadoop 调度器的作用就是将各个 TaskTracker 上的空闲 slot 分配给 Task 使用。slot 分为 Map slot 和 Reduce slot 两种，分别供 MapTask 和 Reduce Task 使用。TaskTracker 通过 slot 数目（可配置参数）限定 Task 的并发度。

Task 分为 Map Task 和 Reduce Task 两种，均由 TaskTracker 启动。HDFS 以 block 块存储数据，MapReduce 处理的最小数据单位为 split。split 如何划分由用户自由设置。split 和 block 之间的关系如图 3-2 所示。

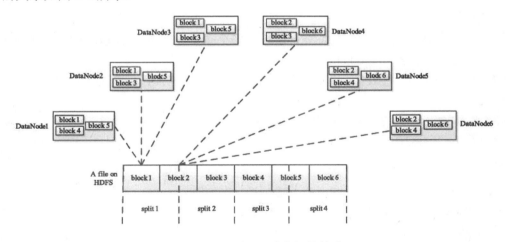

图 3-2　split 和 block 之间的关系

4. HDFS

HDFS 相关内容在 3.2 节和 3.3 节介绍。

3.1.2 MapReduce 的执行步骤

MapReduce 的执行步骤如图 3-3 所示。

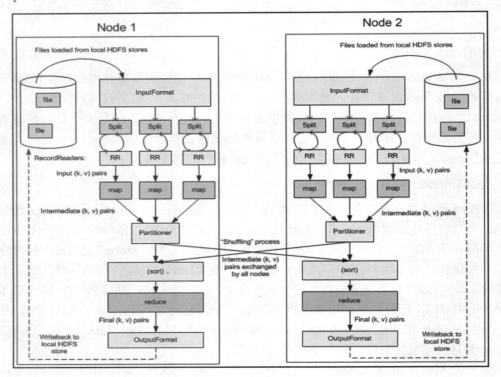

图 3-3 MapReduce 的执行步骤

1. Map 任务处理

（1）读取 HDFS 中的文件。每一行解析成一个<k,v>。每一个键值对调用一次 map 函数。比如：<0,hello you>、<10,hello me>。

（2）覆盖 map()，接收第（1）步产生的<k,v>，进行处理，转换为新的<k,v>输出。比如：<hello,1>、<you,1>、<hello,1>、<me,1>。

（3）对上面（1）（2）两步输出的<k,v>进行分区，默认分为一个区。

（4）对不同分区中的数据进行排序（按照 k）、分组。分组指的是相同 key 的 value 放到一个集合中。

- 排序后：<hello,1>、<hello,1>、<me,1>、<you,1>。
- 分组后：<hello,{1,1}>、<me,{1}>、<you,{1}>。

（5）（可选）对分组后的数据进行归约。

2. Reduce 任务处理

（1）多个 map 任务的输出，按照不同的分区，通过网络 copy 到不同的 reduce 节点上（shuffle）。

（2）对多个 map 的输出进行合并、排序。覆盖 reduce 函数，接收的是分组后的数据，实现自己的业务逻辑，<hello,2> <me,1> <you,1>处理后，产生新的<k,v>输出。

（3）对 reduce 输出的<k,v>写到 HDFS 中。

3.2 HDFS 原理介绍

3.2.1 HDFS 是什么

Hadoop 分布式文件系统（HDFS）被设计成适合运行在通用硬件（commodity hardware）上的分布式文件系统。它和现有的分布式文件系统有很多共同点，同时，它和其他的分布式文件系统的区别也是很明显的。

- HDFS 是一个高度容错性的系统，适合部署在廉价的机器上。
- HDFS 能提供高吞吐量的数据访问，非常适合大规模数据集上的应用。
- HDFS 放宽了一部分 POSIX 约束，来实现流式读取文件系统数据的目的。
- HDFS 在最开始是作为 Apache Nutch 搜索引擎项目的基础架构而开发的。
- HDFS 是 Apache Hadoop Core 项目的一部分。

HDFS 有着高容错性（fault-tolerant）的特点，并且设计用来部署在低廉的（low-cost）硬件上，而且它提供高吞吐量（high throughput）来访问应用程序的数据，适合那些有着超大数据集（large data set）的应用程序。HDFS 放宽了（relax）POSIX 的要求（requirements），这样可以实现流的形式访问（streaming access）文件系统中的数据。

3.2.2 HDFS 架构介绍

1. HDFS 架构

HDFS 架构如图 3-4 所示。

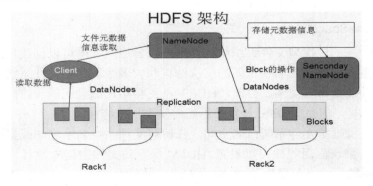

图 3-4　HDFS 架构

从图中可以看出来，HDFS 中三个重要的组件是：

- NameNode
- DataNode
- Sencondary NameNode

2. 数据存储细节

HDFS 的 block 存储如图 3-5 所示。

3. NameNode

NameNode 的目录结构：

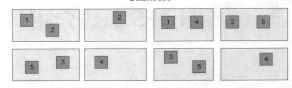

图 3-5　HDFS 的 block 存储

```
${dfs.name.dir}/current /VERSION
                        /edits
                        /fsimage
                        /fstime
```

dfs.name.dir 是 hdfs-site.xml 里配置的若干个目录组成的列表。

NameNode 上保存着 HDFS 的名字空间。对于任何对文件系统元数据产生修改的操作，NameNode 都会使用一种称为 EditLog 的事务日志记录下来。例如，在 HDFS 中创建一个文件，NameNode 就会在 Editlog 中插入一条记录来表示；同样地，修改文件的副本系数也将往 Editlog 插入一条记录。NameNode 在本地操作系统的文件系统中存储这个 Editlog。整个文件系统的名字空间，包括数据块到文件的映射、文件的属性等，都存储在一个称为 fsimage 的文件中，这个文件也是放在 NameNode 所在的本地文件系统上。

NameNode 在内存中保存着整个文件系统的名字空间和文件数据块映射（Blockmap）的映像。这个关键的元数据结构设计得很紧凑，因而一个有 4GB 内存的 NameNode 足够支撑大量的文件和目录。当 NameNode 启动时，它从硬盘中读取 Editlog 和 fsimage，将所有 Editlog 中的事务作用在内存中的 fsimage 上，并将这个新版本的 fsimage 从内存中保存到本地磁盘上，然后删除旧的 Editlog，因为这个旧的 Editlog 的事务都已经作用在 fsimage 上了。这个过程称为一个检查点（checkpoint）。在当前实现中，检查点只发生在 NameNode 启动时，在不久的将来将实现支持周期性的检查点。

4. HDFS NameSpace

HDFS 支持传统的层次型文件组织结构。用户或者应用程序可以创建目录，然后将文件保存在这些目录里。文件系统名字空间的层次结构和大多数现有的文件系统类似：用户可以创建、删除、移动或重命名文件。目前，HDFS 不支持用户磁盘配额和访问权限控制，也不支持硬链接和软链接，但是 HDFS 架构并不妨碍实现这些特性。

NameNode 负责维护文件系统命名空间，任何对文件系统名字空间或属性的修改都将被 NameNode 记录下来。应用程序可以设置 HDFS 保存的文件的副本数目。文件副本的数目称为文件的副本系数，这个信息也是由 NameNode 保存的。

5. DataNode

DataNode 将 HDFS 数据以文件的形式存储在本地的文件系统中，它并不知道有关

HDFS 文件的信息。它把每个 HDFS 数据块存储在本地文件系统的一个单独的文件中。DataNode 并不在同一个目录创建所有的文件，实际上，它用试探的方法来确定每个目录的最佳文件数目，并且在适当的时候创建子目录。在同一个目录中创建所有的本地文件并不是最优的选择，这是因为本地文件系统可能无法高效地在单个目录中支持大量的文件。

当一个 DataNode 启动时，它会扫描本地文件系统，产生一个这些本地文件对应的所有 HDFS 数据块的列表，然后作为报告发送到 NameNode，这个报告就是块状态报告。

6. Secondary NameNode

Secondary NameNode 定期合并 fsimage 和 edits 日志，将 edits 日志文件大小控制在一个限度下。Secondary NameNode 执行步骤如图 3-6 所示。

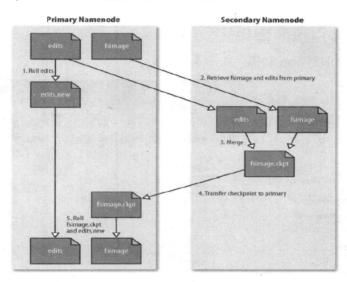

图 3-6 Secondary NameNode 执行步骤

7. Secondary NameNode 处理流程

（1）NameNode 响应 Secondary NameNode 请求，将 edit log 推送给 Secondary NameNode，开始重新写一个新的 edit log。

（2）Secondary NameNode 收到来自 NameNode 的 fsimage 文件和 edit log。

（3）Secondary NameNode 将 fsimage 加载到内存，应用 edit log，并生成一个新的 fsimage 文件。

（4）Secondary NameNode 将新的 fsimage 推送给 NameNode。

（5）NameNode 用新的 fsimage 取代旧的 fsimage，在 fstime 文件中记下检查点发生的时间。

8. HDFS 通信协议

所有的 HDFS 通信协议都是构建在 TCP/IP 协议上。客户端通过一个可配置的端口连接到 NameNode，通过 ClientProtocol 与 NameNode 交互。而 DataNode 是使用 DataNodeProtocol

与 NameNode 交互。在设计上，DataNode 通过周期性地向 NameNode 发送心跳和数据块来保持和 NameNode 的通信，数据块报告的信息包括数据块的属性（即数据块属于哪个文件）、数据块 ID、修改时间等，NameNode 的 DataNode 和数据块的映射关系就是通过系统启动时 DataNode 的数据块报告建立的。从 ClientProtocol 和 DataNodeProtocol 抽象出一个远程过程调用（RPC），在设计上，NameNode 不会主动发起 RPC，而是响应来自客户端和 DataNode 的 RPC 请求。

3.3 HDFS 实战

3.3.1 HDFS 客户端的操作

（1）HDFS 的 FileSystem Shell（如图 3-7 所示）

图 3-7　HDFS 的 FileSystem Shell

（2）启动命令

```
[root@hadoop11 ~]# sh /usr/app/hadoop-2.6.0/sbin/start-dfs.sh
17/03/06 06:10:30 WARN util.NativeCodeLoader: Unable to load native-
hadoop library for your platform... using builtin-java classes where
applicable
Starting namenodes on [hadoop11 hadoop12]
hadoop11: starting namenode, logging to /usr/app/hadoop-
2.6.0/logs/hadoop-root-namenode-hadoop11.out
hadoop12: starting namenode, logging to /usr/app/hadoop-
2.6.0/logs/hadoop-root-namenode-hadoop12.out
hadoop11: starting datanode, logging to /usr/app/hadoop-
2.6.0/logs/hadoop-root-datanode-hadoop11.out
hadoop12: starting datanode, logging to /usr/app/hadoop-
2.6.0/logs/hadoop-root-datanode-hadoop12.out
hadoop13: starting datanode, logging to /usr/app/hadoop-
2.6.0/logs/hadoop-root-datanode-hadoop13.out
Starting journal nodes [hadoop11 hadoop12 hadoop13]
hadoop12: starting journalnode, logging to /usr/app/hadoop-
2.6.0/logs/hadoop-root-journalnode-hadoop12.out
hadoop13: starting journalnode, logging to /usr/app/hadoop-
2.6.0/logs/hadoop-root-journalnode-hadoop13.out
hadoop11: starting journalnode, logging to /usr/app/hadoop-
2.6.0/logs/hadoop-root-journalnode-hadoop11.out
17/03/06 06:10:55 WARN util.NativeCodeLoader: Unable to load native-
hadoop library for your platform... using builtin-java classes where
applicable
Starting ZK Failover Controllers on NN hosts [hadoop11 hadoop12]
hadoop11: starting zkfc, logging to /usr/app/hadoop-
2.6.0/logs/hadoop-root-zkfc-hadoop11.out
hadoop12: starting zkfc, logging to /usr/app/hadoop-
2.6.0/logs/hadoop-root-zkfc-hadoop12.out
```

（3）启动进程

```
[root@hadoop11 ~]# jps
3733 NameNode
3831 DataNode
4013 JournalNode
2682 QuorumPeerMain
4155 DFSZKFailoverController
4262 Jps
```

（4）HDFS 的界面

HDFS 的界面访问地址为 http://hadoop11:50070/dfshealth.html#tab-overview。

HDFS 的界面如图 3-8 所示。

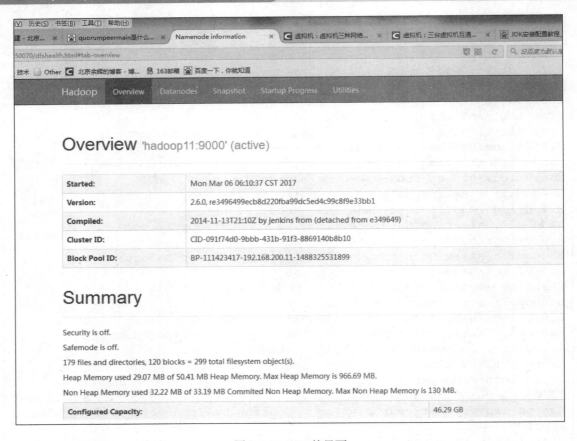

图 3-8　HDFS 的界面

（5）查看目录（HDFS 目录结构如图 3-9 所示）

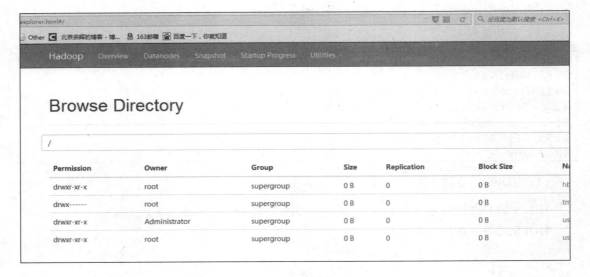

图 3-9　HDFS 目录结构

```
[root@hadoop11 ~]# hadoop fs -ls /
drwxr-xr-x   - root          supergroup          0 2017-03-05 06:14 /hbase
drwx------   - root          supergroup          0 2017-03-03 05:39 /tmp
drwxr-xr-x   - Administrator supergroup          0 2017-03-05 12:06 /user
drwxr-xr-x   - root          supergroup          0 2017-03-03 05:39 /usr
```

（6）增加目录

```
[root@hadoop11 ~]# hadoop fs -ls /usr
drwxr-xr-x   - root supergroup          0 2017-03-05 02:11 /usr/yuhui
```

```
hadoop fs -mkdir /usr/orcale
```

```
[root@hadoop11 ~]# hadoop fs -ls /usr
drwxr-xr-x   - root supergroup          0 2017-03-06 06:22 /usr/orcale
drwxr-xr-x   - root supergroup          0 2017-03-05 02:11 /usr/yuhui
```

（7）删除目录

```
hadoop fs -rm /usr/orcale
```

（8）修改目录

```
hadoop fs -mkdir /usr/orcale
```

```
hadoop fs -mv /usr/orcale /usr/xiaohui
```

（9）上传文件

```
hadoop fs -put hadoop-stop.sh /usr/xiaohui/
```

（10）查看文件

```
hadoop fs -cat /user/xiaohui/hadoop-stop.sh
```

（11）删除文件

```
hadoop fs -rm /usr/xiaohui/hadoop-stop.sh
```

3.3.2 Java 操作 HDFS

1. 注意事项 01

Windows 操作 HDFS 的时候，需要让主机用户 Administrator 有操作权限。HDFS 文件权限如图 3-10 所示。

执行命令如下：

```
hadoop fs -chmod 777 /
```

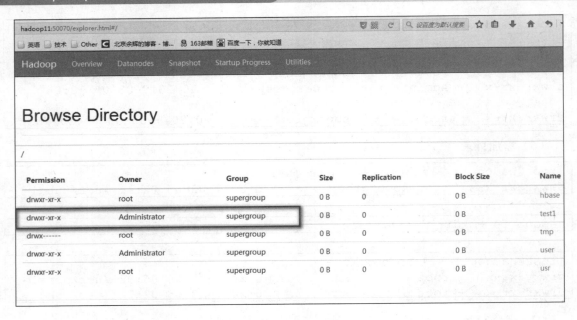

图 3-10　HDFS 文件权限

2. 注意事项 02

将 4 个配置文件 core-site.xml、hdfs-site.xml、mapred-site.xml、yarn-site.xml 放入 resources 中。Java 操作 HDFS 类如图 3-11 所示。

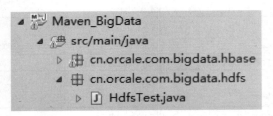

图 3-11　Java 操作 HDFS 类

```java
package cn.orcale.com.bigdata.hdfs;

import java.io.IOException;
import java.io.InputStream;

import org.apache.hadoop.conf.Configuration;
import org.apache.hadoop.fs.FSDataOutputStream;
import org.apache.hadoop.fs.FileStatus;
import org.apache.hadoop.fs.FileSystem;
import org.apache.hadoop.fs.Path;
import org.apache.hadoop.io.IOUtils;

public class HdfsTest {

    FileSystem fs = null;

    // 创建新文件
    public static void createFile(String dst, byte[] contents)
```

```java
        throws IOException {
        Configuration conf = new Configuration();
        FileSystem fs = FileSystem.get(conf);
        Path dstPath = new Path(dst); // 目标路径
        // 打开一个输出流
        FSDataOutputStream outputStream = fs.create(dstPath);
        outputStream.write(contents);
        outputStream.close();
        fs.close();
        System.out.println("文件创建成功!");
    }

    // 上传本地文件
    public static void uploadFile(String src, String dst) throws
IOException {
        Configuration conf = new Configuration();
        FileSystem fs = FileSystem.get(conf);
        Path srcPath = new Path(src); // 原路径
        Path dstPath = new Path(dst); // 目标路径
        // 调用文件系统的文件复制函数,前面参数是指是否删除原文件,true 为删除,默
认为 false
        fs.copyFromLocalFile(false, srcPath, dstPath);

        // 打印文件路径
        System.out.println("Upload to " +
conf.get("fs.default.name"));
        System.out.println("------------list files------------" +
"\n");
        FileStatus[] fileStatus = fs.listStatus(dstPath);
        for (FileStatus file : fileStatus) {
            System.out.println(file.getPath());
        }
        fs.close();
    }

    // 文件重命名
    public static void rename(String oldName, String newName)
            throws IOException {
        Configuration conf = new Configuration();
        FileSystem fs = FileSystem.get(conf);
        Path oldPath = new Path(oldName);
        Path newPath = new Path(newName);
        boolean isok = fs.rename(oldPath, newPath);
        if (isok) {
            System.out.println("rename ok!");
        } else {
            System.out.println("rename failure");
        }
        fs.close();
    }

    // 删除文件
    public static void delete(String filePath) throws IOException {
        Configuration conf = new Configuration();
        FileSystem fs = FileSystem.get(conf);
        Path path = new Path(filePath);
        boolean isok = fs.deleteOnExit(path);
        if (isok) {
            System.out.println("delete ok!");
        } else {
            System.out.println("delete failure");
```

```java
        }
        fs.close();
    }

    // 创建目录
    public static void mkdir(String path) throws IOException {
        Configuration conf = new Configuration();
        FileSystem fs = FileSystem.get(conf);
        Path srcPath = new Path(path);
        boolean isok = fs.mkdirs(srcPath);
        if (isok) {
            System.out.println("create dir ok!");
        } else {
            System.out.println("create dir failure");
        }
        fs.close();
    }

    // 读取文件的内容
    public static void readFile(String filePath) throws IOException {
        Configuration conf = new Configuration();
        FileSystem fs = FileSystem.get(conf);
        Path srcPath = new Path(filePath);
        InputStream in = null;
        try {
            in = fs.open(srcPath);
            IOUtils.copyBytes(in, System.out, 4096, false); // 复制到标准输出流
        } finally {
            IOUtils.closeStream(in);
        }
    }

    public static void main(String[] args) throws IOException {
        /***
         * 本地操作 HDFS 的时候，需要让主机用户 Administrator 有操作权限 hadoop fs -chmod 777 /
         */
        //

        // 测试新建目录
        // mkdir("/test1");

        // 测试上传文件
        // uploadFile("D:\\aaa.txt", "/test1");

        // 测试创建文件
        // byte[] contents = "hello world 世界你好\n".getBytes();
        // createFile("/test1/d.txt",contents);

        // 测试重命名
        // rename("/test1/d.txt", "/test1/dd.txt");

        // 测试读取文件
        // readFile("/test1/aaa.txt");

        // 测试删除文件
        // delete("/test1/dd.txt");  //使用相对路径
```

```
       // delete("/test1");  //删除目录
    }
}
```

3.4 YARN 原理介绍

1. YARN 是什么

Apache Hadoop YARN （Yet Another Resource Negotiator，另一种资源协调者）是一种新的 Hadoop 资源管理器，它是一个通用资源管理系统，可为上层应用提供统一的资源管理和调度，它的引入为集群在利用率、资源统一管理和数据共享等方面带来了巨大好处。

2. YARN 的组成部分

YARN 共有 ResourceManager、NodeManager、JobHistoryServer、Containers、Application Master、job、Task、Client 组成。YARN 的组成示意图如图 3-12 所示。

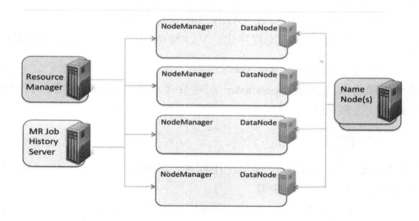

图 3-12　YARN 的组成部分

（1）ResourceManager：一个 Cluster 只有一个，负责资源调度、资源分配等工作。

（2）JobHistoryServer：负责查询 job 运行进度及元数据管理。

（3）NodeManager：运行在 DataNode 节点，负责启动 Application 和对资源的管理。

（4）Containers：Container 通过 ResourceManager 分配，包括容器的 CPU、内存等资源。

（5）ApplicationMaster：通俗来说 ApplicationMaster 相当于包工头，ResourceManager 相当于经理。ResourceManager 首先将任务给 ApplicationMaster，然后 ApplicationMaster 再将 ResourceManager 的指示传达给各个 NodeManager（相当于工人）进行干活。每个 Application 只有一个 ApplicationMaster，运行在 NodeManager 节点，ApplicationMaster 是由 ResourceManager 指派的。

（6）job：一个 Mapper、一个 Reducer 或一个进程的输入列表。job 也可以叫做

Application。

（7）task：一个具体做 Mapper 或 Reducer 的独立的工作单元。task 运行在 NodeManager 的 Container 中。

（8）client：一个提交给 ResourceManager 的 Application 程序。

3. YARN 上的运行应用程序

运行在 YARN 上的应用程序主要分 2 类：

- 短应用程序
- 长应用程序

短应用程序是指一定时间内（可能是秒级、分钟级或小时级，尽管无级别或者更长时间的也存在，但非常少）可运行完成并正常退出的应用程序，比如 MapReduce 作业、Tez DAG 作业等。

长应用程序是指不出意外，永不终止运行的应用程序，通常是一些服务，比如 Storm Service（主要包括 Nimbus 和 Supervisor 两类服务），HBase Service（包括 Hmaster 和 RegionServer 两类服务）等，而它们本身作为一个框架提供了编程接口供用户使用。

尽管这两类应用程序作用不同，一类直接运行数据处理程序，一类用于部署服务（服务之上再运行数据处理程序），但运行在 YARN 上的流程是相同的。

当用户向 YARN 中提交一个应用程序后，YARN 将分两个阶段运行该应用程序：

- 第一个阶段是启动 ApplicationMaster。
- 第二个阶段是由 ApplicationMaster 创建应用程序，为它申请资源，并监控它的整个运行过程，直到运行完成。

4. YARN 的工作流程

（1）步骤 1：用户向 YARN 中提交应用程序，其中包括 ApplicationMaster 程序、启动 ApplicationMaster 的命令、用户程序等。

（2）步骤 2：ResourceManager 为该应用程序分配第一个 Container（这里可以理解为一种资源，比如内存），并与对应的 NodeManager 通信，要求它在这个 Container 中启动应用程序的 ApplicationMaster。

（3）步骤 3：ApplicationMaster 首先向 ResourceManager 注册，这样用户可以直接通过 ResourceManage 查看应用程序的运行状态，然后它将为各个任务申请资源，并监控它的运行状态，直到运行结束，即重复步骤 4~7。

（4）步骤 4：ApplicationMaster 采用轮询的方式通过 RPC 协议向 ResourceManager 申请和领取资源。

（5）步骤 5：一旦 ApplicationMaster 申请到资源后，便与对应的 NodeManager 通信，要求它启动任务。

（6）步骤 6：NodeManager 为任务设置好运行环境（包括环境变量、JAR 包、二进制程序等）后，将任务启动命令写到一个脚本中，并通过运行该脚本启动任务。

（7）步骤 7：各个任务通过某个 RPC 协议向 ApplicationMaster 汇报自己的状态和进

度,以让 ApplicationMaster 随时掌握各个任务的运行状态,从而可以在任务失败时重新启动任务。

在应用程序运行过程中,用户可随时通过 RPC 向 ApplicationMaster 查询应用程序的当前运行状态。

(8)步骤 8:应用程序运行完成后,ApplicationMaster 向 ResourceManager 注销并关闭自己。

Apache YARN 的工作流程如图 3-13 所示。

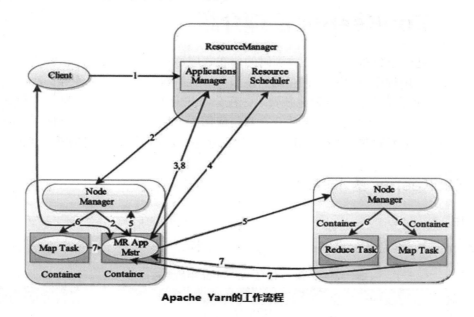

图 3-13 Apache YARN 的工作流程

3.5 小结

前两章已经讲述了 Hadoop 的简介、生态系统、开发环境搭建、集群搭建。从本章开始到第 7 章讲解怎样在 Apache 版本下的 Hadoop 平台进行操作。

本章首先详细描述了 MapReduce 框架结构以及 MapReduce 执行步骤。接着详细描述了 HDFS 的运行原理和框架结构,深入介绍了 HDFS 中每个角色的作用,之后通过 HDFS 客户端和 Eclipse+Java 实现对 HDFS 的具体操作。最后,描述了 YARN 的原理、组成、应用和工作流程。

第 4 章 ZooKeeper 实战

4.1 ZooKeeper 原理介绍

ZooKeeper 是一个分布式的、开放源码的分布式应用程序协调服务，它包含一个简单的原语集，分布式应用程序可以基于它实现同步服务、配置维护和命名服务等。ZooKeeper 是 Hadoop 的一个子项目，其发展历程无须赘述。在分布式应用中，由于工程师不能很好地使用锁机制，以及基于消息的协调机制不适合在某些应用中使用，因此需要有一种可靠的、可扩展的、分布式的、可配置的协调机制来统一系统的状态。ZooKeeper 的目的就在于此。

4.1.1 ZooKeeper 基本概念

1. 角色

ZooKeeper 中的角色主要有以下三类，如表 4-1 所示。

表 4-1 ZooKeeper 中的角色

角色		描述
领导者（Leader）		领导者负责进行投票的发起和决议，更新系统状态
学习者（Learner）	跟随者（Follower）	Follwer 用于接收客户请求并向客户端返回结果，在选主过程中参与投票
	观察者（ObServer）	ObServer 可以接收客户端连接，将写请求转发给 Leader 节点。但 ObServer 不参与投票过程，只同步 leader 的状态。ObServer 的目的是为了扩展系统，提高读取速度
客户端（Client）		请求发起方

ZooKeeper 系统模型如图 4-1 所示。

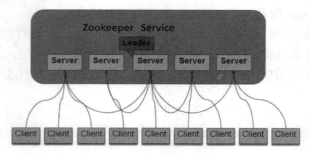

图 4-1 ZooKeeper 系统模型

2. 设计目的

（1）最终一致性：Client 不论连接到哪个 Server，展示给它的都是同一个视图，这是 ZooKeeper 最重要的功能。

（2）可靠性：具有简单、健壮、良好的性能，如果消息 m 被一台服务器接受，那么它将被所有的服务器接受。

（3）实时性：ZooKeeper 保证客户端在一个时间间隔范围内获得服务器的更新信息，或者服务器失效的信息。但由于网络延时等原因，ZooKeeper 不能保证两个客户端能同时得到刚刚更新的数据。如果需要最新数据，应该在读数据之前调用 sync()接口。

（4）等待无关（wait-free）：慢的或者失效的 Client 不得干预快速的 Client 的请求，使得每个 Client 都能有效地等待。

（5）原子性：更新只能成功或者失败，没有中间状态。

（6）顺序性：包括全局有序和偏序两种顺序。全局有序是指如果在一台服务器上消息 a 在消息 b 前发布，则在所有 Server 上消息 a 都将在消息 b 前被发布；偏序是指如果一个消息 b 在消息 a 后被同一个发送者发布，a 必将排在 b 前面。

4.1.2 ZooKeeper 工作原理

ZooKeeper 的核心是原子广播，这个机制保证了各个 Server 之间的同步。实现这个机制的协议叫做 Zab 协议。Zab 协议有两种模式，它们分别是恢复模式（选主）和广播模式（同步）。当服务启动或者在领导者崩溃后，Zab 就进入了恢复模式，当领导者被选举出来，且大多数 Server 完成了和 Leader 的状态同步以后，恢复模式就结束了。状态同步保证了 Leader 和 Server 具有相同的系统状态。

为了保证事务的顺序一致性，ZooKeeper 采用了递增的事务 id 号（zxid）来标识事务。所有的提议（proposal）都在被提出的时候加上了 zxid。实现中 zxid 是一个 64 位的数字，高 32 位是 epoch 用来标识 Leader 关系是否改变，每次一个 Leader 被选出来，它都会有一个新的 epoch，标识当前属于那个 Leader 的统治时期；低 32 位用于递增计数。

每个 Server 在工作过程中有三种状态：

- LOOKING：当前 Server 不知道 Leader 是谁，正在搜寻。
- LEADING：当前 Server 即为选举出来的 Leader。
- FOLLOWING：Leader 已经选举出来，当前 Server 与之同步。

1. 选主流程

当 Leader 崩溃或者 Leader 失去大多数的 Follower，这时候 zk 进入恢复模式，恢复模式需要重新选举出一个新的 Leader，让所有的 Server 都恢复到一个正确的状态。zk 的选举算法有两种：一种是基于 basic paxos 实现的，另外一种是基于 fast paxos 算法实现的。系统默认的选举算法为 fast paxos。先介绍 basic paxos 流程：

（1）选举线程由当前 Server 发起选举的线程担任，其主要功能是对投票结果进行统计，并选出推荐的 Server。

（2）选举线程首先向所有 Server 发起一次询问（包括自己）。

（3）选举线程收到回复后，验证是否是自己发起的询问（验证 zxid 是否一致），然后获取对方的 id(myid)，并存储到当前询问对象列表中，最后获取对方提议的 leader 相关信息（id、zxid），并将这些信息存储到当次选举的投票记录表中。

（4）收到所有 Server 回复以后，就计算出 zxid 最大的那个 Server，并将这个 Server 相关信息设置成下一次要投票的 Server。

（5）线程将当前 zxid 最大的 Server 设置为当前 Server 要推荐的 Leader，如果此时获胜的 Server 获得 n/2 + 1 的 Server 票数，设置当前推荐的 Leader 为获胜的 Server，将根据获胜的 Server 相关信息设置自己的状态，否则，继续这个过程，直到 leader 被选举出来。

通过上面的流程分析我们可以得出：要使 Leader 获得多数 Server 的支持，则 Server 总数必须是奇数 2n+1，且存活的 Server 的数目不得少于 n+1。

每个 Server 启动后都会重复以上流程。在恢复模式下，如果是刚从崩溃状态恢复的或者刚启动的 Server，还会从磁盘快照中恢复数据和会话信息，zk 会记录事务日志并定期进行快照，方便状态恢复。选主流程如图 4-2 所示。

fast paxos 流程是在选举过程中，某 Server 首先向所有 Server 提议自己要成为 leader，当其他 Server 收到提议以后，解决 epoch 和 zxid 的冲突，并接受对方的提议，然后向对方发送接受提议完成的消息，重复这个流程，最后一定能选举出 Leader。fast paxos 流程如图 4-3 所示。

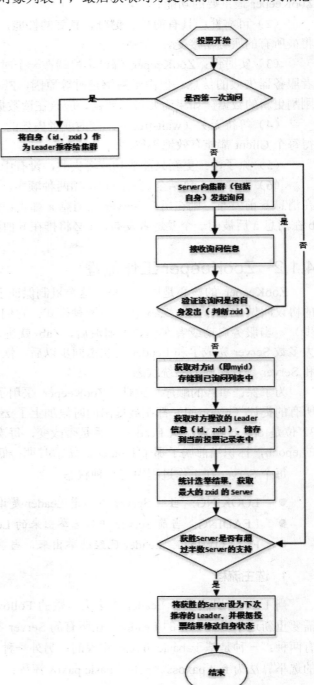

图 4-2　选主流程

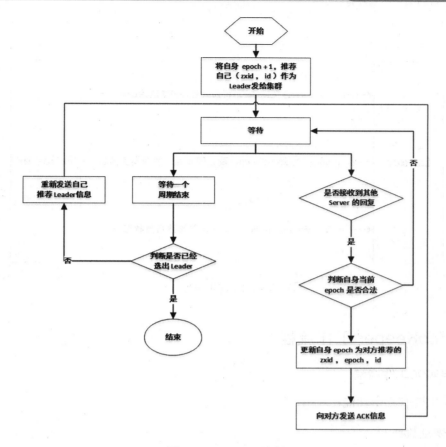

图 4-3　fast paxos 流程

2. 同步流程

选完 leader 以后，zk 就进入状态同步过程。

- leader 等待 Server 连接。
- Follower 连接 leader，将最大的 zxid 发送给 Leader。
- Leader 根据 Follower 的 zxid 确定同步点。
- 完成同步后通知 Follower 已经成为 uptodate 状态。
- Follower 收到 uptodate 消息后，又可以重新接受 Client 的请求进行服务了。

同步流程如图 4-4 所示。

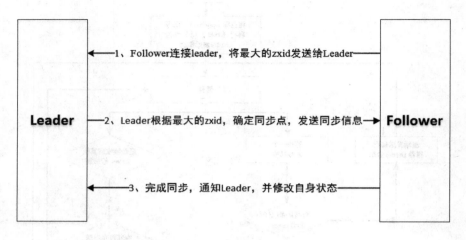

图 4-4 同步流程

4.1.3 ZooKeeper 工作流程

1. Leader 工作流程

Leader 主要有三个功能：

（1）恢复数据。

（2）维持与 Learner 的心跳，接收 Learner 请求并判断 Learner 的请求消息类型。

（3）Learner 的消息类型主要有 PING 消息、REQUEST（request）消息、ACK 消息、REVALIDATE 消息，根据不同的消息类型，进行不同的处理。

- PING 消息是指 Learner 的心跳信息。
- REQUEST 消息是 Follower 发送的提议信息，包括写请求及同步请求。
- ACK 消息是 Follower 对提议的回复，超过半数的 Follower 通过，则 commit 该提议。
- REVALIDATE 消息是用来延长 SESSION 有效时间。

Leader 的工作流程简图如图 4-5 所示。在实际实现中，流程要比下图复杂得多，启动了三个线程来实现功能。

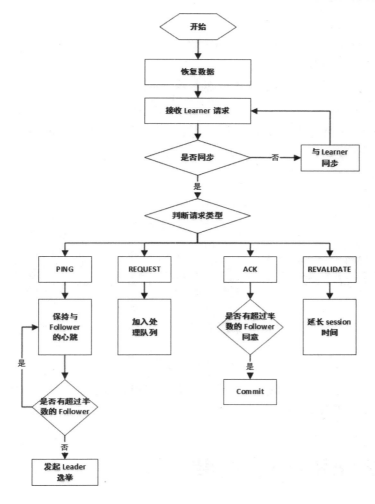

图 4-5　Leader 的工作流程简图

2. Follower 工作流程

Follower 主要有 4 个功能：

- 向 Leader 发送请求（PING 消息、REQUEST 消息、ACK 消息、REVALIDATE 消息）。
- 接收 Leader 消息并进行处理。
- 接收 Client 的请求，如果为写请求，发送给 Leader 进行投票。
- 返回 Client 结果。

Follower 的消息循环处理如下几种来自 Leader 的消息：

- PING 消息：心跳消息。
- PROPOSAL 消息：Leader 发起的提案，要求 Follower 投票。
- COMMIT 消息：服务器端最新一次提案的信息。

- UPTODATE 消息：表明同步完成。
- REVALIDATE 消息：根据 Leader 的 REVALIDATE 结果，关闭待 revalidate 的 session 还是允许其接受消息。
- SYNC 消息：返回 SYNC 结果到客户端，这个消息最初由客户端发起，用来强制得到最新的更新。

Follower 的工作流程简图如图 4-6 所示。在实际实现中，Follower 是通过 5 个线程来实现功能的。

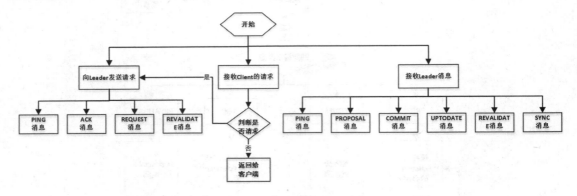

图 4-6　Follower 的工作流程简图

对于 observer 的流程不再叙述，observer 流程和 Follower 的唯一不同的地方就是 observer 不会参加 Leader 发起的投票。

4.2　ZooKeeper 安装

1. ZooKeeper 安装

第 2 章"Hadoop 集群搭建"中已经介绍了 ZooKeeper 的安装步骤和启动步骤。

2. 批量启动 ZooKeeper

下面是一个批量启动 ZooKeeper 的脚本，如图 4-7 所示。它用于避免读者每次登录所有机器单独启动 ZooKeeper。主要原理是远程发送命令，从 A 服务器到达 B 服务器，启动某条 shell 命令。

```
[root@hadoop11 ~]# sh all-zookeeper-start.sh
JMX enabled by default
Using config: /usr/app/zookeeper-3.4.6/bin/../conf/zoo.cfg
Starting zookeeper ... STARTED
JMX enabled by default
Using config: /usr/app/zookeeper-3.4.6/bin/../conf/zoo.cfg
Starting zookeeper ... STARTED
JMX enabled by default
Using config: /usr/app/zookeeper-3.4.6/bin/../conf/zoo.cfg
Starting zookeeper ... STARTED
```

图 4-7　批量启动 ZooKeeper 的脚本

all-zookeeper-start.sh

```
#!/bin/sh
ssh root@hadoop11 "bash" < /root/zookeeper-start.sh
ssh root@hadoop12 "bash" < /root/zookeeper-start.sh
ssh root@hadoop13 "bash" < /root/zookeeper-start.sh
```

zookeeper-start.sh

```
#!/bin/sh
/usr/app/zookeeper-3.4.6/bin/./zkServer.sh start
```

all-zookeeper-stop.sh

```
#!/bin/sh
ssh root@hadoop11 "bash" < /root/zookeeper-stop.sh
ssh root@hadoop12 "bash" < /root/zookeeper-stop.sh
ssh root@hadoop13 "bash" < /root/zookeeper-stop.sh
```

zookeeper-stop.sh

```
#!/bin/sh
/usr/app/zookeeper-3.4.6/bin/./zkServer.sh stop
```

3. ZooKeeper 进程

```
[root@hadoop11 ~]# jps
2682 QuorumPeerMain
2879 Jps
[root@hadoop11 ~]#
```

4. 查看 ZooKeeper 状态

使用命令：/usr/app/zookeeper-3.4.6/bin/./zkServer.sh status 查看，三台节点 ZooKeeper 的状态如图 4-8 所示。

```
[root@hadoop11 ~]# /usr/app/zookeeper-3.4.6/bin/./zkServer.sh status
JMX enabled by default
Using config: /usr/app/zookeeper-3.4.6/bin/../conf/zoo.cfg
Mode: follower
[root@hadoop12 ~]# /usr/app/zookeeper-3.4.6/bin/./zkServer.sh status
JMX enabled by default
Using config: /usr/app/zookeeper-3.4.6/bin/../conf/zoo.cfg
Mode: leader
[root@hadoop12 ~]#
[root@hadoop13 ~]# /usr/app/zookeeper-3.4.6/bin/./zkServer.sh status
JMX enabled by default
Using config: /usr/app/zookeeper-3.4.6/bin/../conf/zoo.cfg
Mode: follower
[root@hadoop13 ~]#
```

图 4-8 三台节点 ZooKeeper 的状态

4.3 ZooKeeper 实战

4.3.1 ZooKeeper 客户端的操作

（1）启动

记住：操作时候查看同步情况。

启动命令：/usr/app/zookeeper-3.4.6/bin/zkCli.sh -server localhost:2181。ZooKeeper 下 bin 目录结构如图 4-9 所示。

```
[root@hadoop11 bin]# ls
README.txt    zkCli.cmd    zkEnv.cmd    zkServer.cmd    zookeeper.out
zkCleanup.sh  zkCli.sh     zkEnv.sh     zkServer.sh
[root@hadoop11 bin]# bash zkCli.sh -server localhost:2181
```

图 4-9 ZooKeeper 下 bin 目录结构

（2）帮助

```
[zk: localhost:2181(CONNECTED) 0] help
ZooKeeper -server host:port cmd args
    connect host:port
    get path [watch]
    ls path [watch]
    set path data [version]
    rmr path
    delquota [-n|-b] path
    quit
    printwatches on|off
    create [-s] [-e] path data acl
    stat path [watch]
    close
    ls2 path [watch]
    history
    listquota path
    setAcl path acl
    getAcl path
    sync path
    redo cmdno
    addauth scheme auth
    delete path [version]
    setquota -n|-b val path
```

（3）查看目录

```
[zk: localhost:2181(CONNECTED) 1] ls /
[hadoop, hbase, root, xiaohui, hadoop-ha, zookeeper, yuhui, yarn-leader-election]
```

（4）创建目录

```
[zk: localhost:2181(CONNECTED) 2] create /zk_test my_data
Created /zk_test
```

(5) 获取数据

```
[zk: localhost:2181(CONNECTED) 4] get /zk_test
my_data
cZxid = 0xa00000002
ctime = Mon Mar 06 05:54:57 CST 2017
mZxid = 0xa00000002
mtime = Mon Mar 06 05:54:57 CST 2017
pZxid = 0xa00000002
cversion = 0
dataVersion = 0
aclVersion = 0
ephemeralOwner = 0x0
dataLength = 7
numChildren = 0
```

(6) 设置数据

```
[zk: localhost:2181(CONNECTED) 5] set /zk_test junk
cZxid = 0xa00000002
ctime = Mon Mar 06 05:54:57 CST 2017
mZxid = 0xa00000003
mtime = Mon Mar 06 05:55:47 CST 2017
pZxid = 0xa00000002
cversion = 0
dataVersion = 1
aclVersion = 0
ephemeralOwner = 0x0
dataLength = 4
numChildren = 0
```

(7) 删除路径

```
[zk: localhost:2181(CONNECTED) 7] delete /zk_test
```

4.3.2 Java 操作 ZooKeeper

Java 操作 ZooKeeper 类和目录如图 4-10 所示。

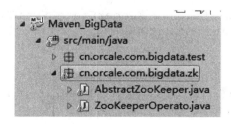

图 4-10 Java 操作 ZooKeeper 类和目录

AbstractZooKeeper 类实现：

```
package cn.orcale.com.bigdata.zk;
```

```java
import java.io.IOException;
import java.util.concurrent.CountDownLatch;

import org.apache.commons.logging.Log;
import org.apache.commons.logging.LogFactory;
import org.apache.zookeeper.WatchedEvent;
import org.apache.zookeeper.Watcher;
import org.apache.zookeeper.ZooKeeper;
import org.apache.zookeeper.Watcher.Event.KeeperState;

public class AbstractZooKeeper implements Watcher {
    private static Log log =
LogFactory.getLog(AbstractZooKeeper.class
            .getName());

    // 缓存时间
    private static final int SESSION_TIME = 2000;
    protected ZooKeeper zooKeeper;
    protected CountDownLatch countDownLatch = new CountDownLatch(1);

    public void connect(String hosts) throws IOException,
InterruptedException {
        zooKeeper = new ZooKeeper(hosts, SESSION_TIME, this);
        countDownLatch.await();
    }

    /*
     * (non-Javadoc)
     *
     * @see
     *
org.apache.zookeeper.Watcher#process(org.apache.zookeeper.WatchedEvent)
     */
    public void process(WatchedEvent event) {
        // TODO Auto-generated method stub
        if (event.getState() == KeeperState.SyncConnected) {
            countDownLatch.countDown();
        }
    }

    public void close() throws InterruptedException {
        zooKeeper.close();
    }
}
```

ZooKeeperOperato 类实现：

```
package cn.orcale.com.bigdata.zk;
```

```java
import java.util.List;

import org.apache.commons.logging.Log;
import org.apache.commons.logging.LogFactory;
import org.apache.zookeeper.CreateMode;
import org.apache.zookeeper.KeeperException;
import org.apache.zookeeper.ZooDefs.Ids;
import org.apache.zookeeper.data.Stat;

/***
 *
 *
 * @author yuhui
 * @date 2017年3月1日 下午9:10:23
 */
public class ZooKeeperOperato extends AbstractZooKeeper {

    private static Log log = LogFactory
            .getLog(ZooKeeperOperato.class.getName());

    // 创建目录
    public void create(String path, byte[] data) throws KeeperException,
            InterruptedException {

        Stat exists = this.pathExists(path);
        if (exists == null) {
            this.zooKeeper.create(path, data, Ids.OPEN_ACL_UNSAFE,
                    CreateMode.PERSISTENT);
            System.out.println(path + " 创建成功");
        }

    }

    // 检测目录
    public Stat pathExists(String Path) throws KeeperException,
            InterruptedException {
        Stat stat = null;
        stat = zooKeeper.exists(Path, false);
        return stat;

    }

    // 获取子目录
    public void getChild(String path) throws KeeperException,
            InterruptedException {
        try {
            List<String> list = this.zooKeeper.getChildren(path, false);
```

```java
            if (list.isEmpty()) {
                System.out.println(path + "中没有节点");
            } else {
                System.out.println(path + "中存在节点");
                for (String child : list) {
                    System.out.println("节点为: " + child);
                }
            }
        } catch (KeeperException.NoNodeException e) {
            // TODO: handle exception
            throw e;

        }
    }

    // 获取数据
    public byte[] getData(String path) throws KeeperException,
            InterruptedException {
        return this.zooKeeper.getData(path, false, null);
    }

    // 删除目录
    public void deletePath(String path) throws KeeperException,
            InterruptedException {

        zooKeeper.delete(path, -1);
        System.out.println(path + "删除成功");
    }

    public static void main(String[] args) {

        try {
            //hadoop11为连接 ZooKeeper 主机
            ZooKeeperOperato zkoperator = new ZooKeeperOperato();

            zkoperator.connect("hadoop11");

            byte[] data = new byte[] { 'a', 'b', 'c', 'd' };

            // zkoperator.create("/xiaohui",null);
            // zkoperator.deletePath("/xiaohui");

            // zkoperator.create("/xiaohui/child1",data);
            //
System.out.println((Arrays.toString(zkoperator.getData("/xiaohui/child1"))));
            //
            // zkoperator.create("/xiaohui/child2",data);
            //
System.out.println(Arrays.toString(zkoperator.getData("/xiaohui/child2")))
```

```
;
            // System.out.println("节点孩子信息:");
            // zkoperator.getChild("/xiaohui");

            zkoperator.close();

        } catch (Exception e) {
            e.printStackTrace();
        }

    }
}
```

4.3.3 Scala 操作 ZooKeeper

```
package com.ou.cn.utils


import kafka.common.TopicAndPartition
import org.apache.curator.framework.CuratorFrameworkFactory
import org.apache.curator.retry.ExponentialBackoffRetry
import org.slf4j.LoggerFactory

import scala.collection.JavaConversions._

/***
 *
 *
 * @author yuhui
 * @date 2017年3月1日 下午9:10:23
 */

object ZookeeperHelper {

  val ZOOKEEPER_CONNECT="hadoop11:2181, hadoop12:2181, hadoop13:2181"

  val LOG = LoggerFactory.getLogger(ZookeeperHelper.getClass)
  val client = {
    val client = CuratorFrameworkFactory
      .builder
      .connectString(ZOOKEEPER_CONNECT)
      .retryPolicy(new ExponentialBackoffRetry(1000, 3))
      .namespace("ou")
      .build()
    client.start()
    client
  }

  //zookeeper 创建路径
  def ensurePathExists(path: String): Unit = {
```

```scala
    if (client.checkExists().forPath(path) == null) {
      client.create().creatingParentsIfNeeded().forPath(path)
    }
  }

  //zookeeper 加载 offset 的方法
  def loadOffsets(topicSet: Set[String], defaultOffset:
Map[TopicAndPartition, Long]): Map[TopicAndPartition, Long] = {
    val kafkaOffsetPath = s"/kafkaOffsets"
    ensurePathExists(kafkaOffsetPath)
    val offsets = for {
    //t 就是路径 webstatistic/kafkaOffsets 下面的子目录遍历
      t <- client.getChildren.forPath(kafkaOffsetPath)
      if topicSet.contains(t)
    //p 就是新路径   /webstatistic/kafkaOffsets/donews_website
      p <- client.getChildren.forPath(s"$kafkaOffsetPath/$t")
    } yield {
    //遍历路径下面的 partition 中的 offset
      val data = client.getData.forPath(s"$kafkaOffsetPath/$t/$p")
    //将 data 变成 Long 类型
      val offset = java.lang.Long.valueOf(new String(data)).toLong
      (TopicAndPartition(t, Integer.parseInt(p)), offset)
    }
    defaultOffset ++ offsets.toMap
  }

  //zookeeper 存储 offset 的方法
  def storeOffsets(offsets: Map[TopicAndPartition, Long]): Unit = {
    val kafkaOffsetPath = s"/kafkaOffsets"
    if (client.checkExists().forPath(kafkaOffsetPath) == null) {
client.create().creatingParentsIfNeeded().forPath(kafkaOffsetPath)
    }
    for ((tp, offset) <- offsets) {
      val data = String.valueOf(offset).getBytes
      val path = s"$kafkaOffsetPath/${tp.topic}/${tp.partition}"
      ensurePathExists(path)
      client.setData().forPath(path, data)
    }
  }

  def main(args: Array[String]) {

    //获取到 namespace
    println(client.getNamespace)

    //创建路径
    val kafkaOffsetPath = "/kafkaOffsets"
    if (client.checkExists().forPath(kafkaOffsetPath) == null) {
```

```
client.create().creatingParentsIfNeeded().forPath(kafkaOffsetPath)
  }
  //删除路径
  client.delete().forPath("/kafkaOffsets/web/1")

  //存储值
  val offsets : Map[TopicAndPartition, Long] =
Map(TopicAndPartition("web",1) ->4444, TopicAndPartition("web",2)-
>2222 )
  storeOffsets(offsets)

  //获取值
  val topicSet = Set("web")
  val offsetstoMap:Map[TopicAndPartition, Long]=
loadOffsets(topicSet,Map(TopicAndPartition("web",1) ->0))
  offsetstoMap.keySet.foreach(line=>{
    val topicName = line.topic
    val topicPartition = line.partition
    val data =
client.getData.forPath(s"$kafkaOffsetPath/$topicName/$topicPartition")
    val offset = java.lang.Long.valueOf(new String(data)).toLong
    println("路径"+s"$kafkaOffsetPath/$topicName/$topicPartition"+"
的值为:"+ offset)
  })

  }
}
```

4.4 小结

ZooKeeper 是一个为分布式应用提供一致性服务的软件。本章讲解 ZooKeeper 组件的原理和实战。原理方面详细描述了 ZooKeeper 的基本概念、ZooKeeper 工作原理、ZooKeeper 工作流程。实战方面通过 ZooKeeper 客户端、Eclipse+Java、Intellij IDEA+Scala 对 ZooKeeper 进行操作。

第 5 章

◀ MapReduce实战 ▶

5.1 前期准备

1. HDFS 路径准备

在 HDFS 建立目录，并将处理文件上传上去。日志在 HDFS 展示如图 5-1 所示。

```
hadoop fs -mkdir /usr
hadoop fs -mkdir /usr/xiaohui
hadoop fs -mkdir /usr/xiaohui/mapreduce
hadoop fs -mkdir /usr/xiaohui/mapreduce/data
hadoop fs -put NginxData.txt /usr/xiaohui/mapreduce/data/
```

Permission	Owner	Group	Size	Replication	Block Size	Name
-rw-r--r--	root	supergroup	19.58 KB	3	128 MB	NginxData.txt

/usr/xiaohui/mapreduce/data

图 5-1　日志在 HDFS 展示

2. NginxData 日志

```
1436992033 123.150.156.4  d77dbbbd035355347be6c52305d89281
    c007235244ad9d8edd3d8f082ff9ed3d    si2.mfniu.com  1920    1080    32
    zh-CN
    http://si2.mfniu.com/HTML_Content_Cache/stock_indicator/1/601727.
html    http://si2.mfniu.com/    Mozilla/5.0 (Windows NT 6.1)
AppleWebKit/537.36 (KHTML, like Gecko) Chrome/31.0.1650.63
Safari/537.36 huatuozhenguseo    _trackPageview    si2.mfniu.com
    10119
1436993140 110.84.190.37  f0171969068dc342029323ca9bc42c10
    8b631869a0f53c227600859deb0bb777    si.mfniu.com   1920    1080    32
        http://si.mfniu.com/Default.aspx
    http://si.mfniu.com/Hot.aspx    Mozilla/5.0 (compatible; MSIE
9.0; Windows NT 6.1; WOW64; Trident/5.0)   huatuozhengu
    _trackPageview        si.mfniu.com    10119
1436991363 113.109.196.169    570116d5cdd70440b30f46a0f5dc43a3
```

```
         2dc416e0eb9673be163853e01b4b6eb9    si.mfniu.com    1600    900 32
         http://si.mfniu.com/HTML_Content_Cache/stock_indicator/1/600239.h
tml http://si.mfniu.com/choice.aspx Mozilla/4.0 (compatible; MSIE
7.0; Windows NT 6.1; WOW64; Trident/7.0; SLCC2; .NET CLR
2.0.50727; .NET CLR 3.5.30729; .NET CLR
3.0.30729; .NET4.0C; .NET4.0E) huatuozhengu    _trackPageview
         si.mfniu.com    10119
1436993038 112.90.231.5    116be797dfd7d207f1df0ea832326da5
         7f54d2cea4bc2c6b93f6eaed3a1ccca3    www.88mf.com    1280    720 32
         zh-CN    http://www.88mf.com/hugangtong/9605.html
         http://www.baidu.com/link?url=SiqiWHX8eIj7pASO1IPK97gKx8Xf1zviJEc
uPxXZzI4Rk0ipTN81bMAHokVXVvvpvvZPQxuf-
AionY1C60eF0_&wd=&eqid=976d0ed800028ab90000000355a6c5ff Mozilla/5.0
(Windows NT 6.1) AppleWebKit/537.36 (KHTML, like Gecko)
Chrome/31.0.1650.63 Safari/537.36    fuhongsheng    _trackPageview
         www.baidu.com 10057
1436991538 112.67.214.106    f6a1c849250e4f4b341f7fb868d2df81
         30e65edd1471985b3766561cd8d87363    www.88mf.com    1366    768 32
         zh-CN    http://www.88mf.com/chaoguruanjian/813.html
         https://www.baidu.com/s?ie=utf-
8&f=8&rsv_bp=1&tn=site888_3_pg&wd=%E8%82%A1%E7%A5%A8%E8%A1%8C%E6%83%
85%E4%B8%AD%E7%9A%84%E8%82%A1%E6%9C%AC%E6%98%AF%E4%BB%80%E4%B9%88%E6
%84%8F%E6%80%9D&rsv_pq=e4adfc7e0000bf28&rsv_t=41c9hHVDgRIOTrel4yteuZ
51F1jIYDrrN1xymVUeyT1QW%2Fvpp8O%2B7w3ok0WcX5qLQFwQ&rsv_enter=1&rsv_s
ug3=18&bs=%E8%82%A1%E7%A5%A8%E8%A1%8C%E6%83%85%E4%B8%AD%E7%9A%84%E8%
82%A1%E6%9C%AC%E6%98%AF%E4%BB%80%E4%B9%88 Mozilla/5.0 (Windows NT
6.1) AppleWebKit/537.36 (KHTML, like Gecko) Chrome/31.0.1650.63
Safari/537.36 fuhongsheng    _trackPageview        www.baidu.com 10057
1436993364 1.82.184.71    486c758b3c2a3dca05587149919a6d51
         89d341d7c550225568a6132601a4e673    si.mfniu.com    1366    768 32
         http://si.mfniu.com/hot.aspx
         http://si.mfniu.com/HTML_Content_Cache/stock_indicator/1/600881.h
tml Mozilla/4.0 (compatible; MSIE 7.0; Windows NT 6.1; WOW64;
Trident/4.0; SLCC2; .NET CLR 2.0.50727; .NET CLR 3.5.30729; .NET CLR
3.0.30729; .NET4.0C; .NET4.0E; Media Center PC 6.0; InfoPath.2)
         huatuozhengu    _trackPageview        si.mfniu.com    10119
1436993503 1.82.184.71    20eb96e05672012b76a0a470a6bab23b
         fabeb5633773eb6b19b59a31d5dd2d5e    si.mfniu.com    1366    768 32
         http://si.mfniu.com/choice.aspx http://si.mfniu.com/hot.aspx
         Mozilla/4.0 (compatible; MSIE 7.0; Windows NT 6.1; WOW64;
Trident/4.0; SLCC2; .NET CLR 2.0.50727; .NET CLR 3.5.30729; .NET CLR
3.0.30729; .NET4.0C; .NET4.0E; Media Center PC 6.0; InfoPath.2)
         huatuozhengu    _trackPageview        si.mfniu.com    10119
1436993654 123.91.75.131 aab18ceb87585548f363e793c2130f90
         6a051d86144237ca1cc2a3efeccda960    si.mfniu.com    1680    1050    32
             http://si.mfniu.com/choice.aspx
         http://si.mfniu.com/?userid=269903&sessionid=34e5cbf0t1la4ejj180i
q24h23 Mozilla/4.0 (compatible; MSIE 8.0; Windows NT 6.1; WOW64;
Trident/4.0; SLCC2; .NET CLR 2.0.50727; .NET CLR 3.5.30729; .NET CLR
3.0.30729; Media Center PC 6.0) huatuozhengu    _trackPageview
```

```
        si.mfniu.com    10119
1436991345 46.189.73.173 29796b238595d5aef1564f069f09fb01
    57e3f62eeaa791c6d7f3c1b29c05f228    www.88mf.com    1280    720 24
    zh-CN    http://www.88mf.com/monichaogu/1757.html
    https://www.google.com.hk/    Mozilla/5.0 (Windows NT 6.1)
AppleWebKit/537.36 (KHTML, like Gecko) Chrome/43.0.2357.134
Safari/537.36 fuhongsheng    _trackPageview        www.google.com.hk
    10111
1436993124 110.84.190.37 9b694e5cd0c5a5831ff666390c451083
    a360f3783ff66e4303f1f3613f98772c    si.mfniu.com    1920    1080    32
    http://si.mfniu.com/Hot.aspx
    http://si.mfniu.com/choice.aspx Mozilla/5.0 (compatible; MSIE
9.0; Windows NT 6.1; WOW64; Trident/5.0)    huatuozhengu
    _trackPageview        si.mfniu.com    10119
1436993511 1.82.184.71    ff40752455be6f7c0279c0c373254ad3
    84382d99758363255bde877b32ca59e5    si.mfniu.com    1366    768 32
    http://si.mfniu.com/Default.aspx
    http://si.mfniu.com/choice.aspx Mozilla/4.0 (compatible; MSIE
7.0; Windows NT 6.1; WOW64; Trident/4.0; SLCC2; .NET CLR
2.0.50727; .NET CLR 3.5.30729; .NET CLR
3.0.30729; .NET4.0C; .NET4.0E; Media Center PC 6.0; InfoPath.2)
    huatuozhengu    _trackPageview        si.mfniu.com    10119
1436993121 110.84.190.37 8d9371f8834709cc8c347f751c0f3bef
    b287584e4821105f52f1d0aaf1a8681e    si.mfniu.com    1920    1080    32
        http://si.mfniu.com/choice.aspx
    http://si.mfniu.com/HTML_Content_Cache/stock_indicator/1/000816.h
tml Mozilla/5.0 (compatible; MSIE 9.0; Windows NT 6.1; WOW64;
Trident/5.0)    huatuozhengu    _trackPageview        si.mfniu.com    10119
1436991332 113.109.196.169    570116d5cdd70440b30f46a0f5dc43a3
    2dc416e0eb9673be163853e01b4b6eb9    si.mfniu.com    1600    900 32
    http://si.mfniu.com/choice.aspx
    http://si.mfniu.com/HTML_Content_Cache/stock_indicator/1/600893.h
tml Mozilla/4.0 (compatible; MSIE 7.0; Windows NT 6.1; WOW64;
Trident/7.0; SLCC2; .NET CLR 2.0.50727; .NET CLR 3.5.30729; .NET CLR
3.0.30729; .NET4.0C; .NET4.0E) huatuozhengu    _trackPageview
    si.mfniu.com    10119
1436990555 120.210.170.3 436de9695fb756efd9503d5627baad38
    ee6502008a50f161bc79168602bdd731    www.88mf.com    1366    768 32
    zh-CN    http://www.88mf.com/monichaogu/1958.html
    http://www.baidu.com/link?url=QLxMiCTQdLgOpdH8I0BjfdsW2laDn3sroGr
WhcDOrP_uITzwc4WT21IMtw9cpKFIw2CM-
KjBy_zPwkDIfKKaGK&wd=&eqid=e5dd3d030001e6aa0000000355a6bc4d
    Mozilla/5.0 (Windows NT 6.1) AppleWebKit/537.36 (KHTML, like
Gecko) Chrome/31.0.1650.63 Safari/537.36    fuhongsheng
    _trackPageview        www.baidu.com 10057
1436993762 123.91.75.131    7ce7142e24c08fee8c24a470d85f1a4e
    3043ce98111e92a67aaa7540c2d7d6b9    si.mfniu.com    1680    1050    32
        http://si.mfniu.com/choice.aspx
    http://si.mfniu.com/HTML_Content_Cache/stock_indicator/1/000025.h
tml Mozilla/4.0 (compatible; MSIE 8.0; Windows NT 6.1; WOW64;
```

```
Trident/4.0; SLCC2; .NET CLR 2.0.50727; .NET CLR 3.5.30729; .NET CLR
3.0.30729; Media Center PC 6.0)    huatuozhengu    _trackPageview
    si.mfniu.com    10119
1436993101  58.35.139.22    b481af465fdfbab14d9135ec3300ed1f
    76c6c7440526452c41dbd2118715ec7c    www.88mf.com    768  1024    32
    zh-cn    http://www.88mf.com/heimajiangu/22509.html
    https://www.baidu.com/s?tn=SE_baiduhomet2_isv1k3dg&rsv_idx=3&ie=u
tf-
8&bs=%E5%92%8C%E5%90%9B%E5%95%86%E5%AD%A6%E9%99%A2&dsp=ipad&f=8&rsv_
bp=1&wd=%E5%92%8C%E5%90%9B%E5%95%86%E5%AD%A6%E9%99%A2+%E6%B1%87%E5%8
6%A0&rsv_sug3=14&rsv_sug4=806&rsv_sug1=8&inputT=13824    Mozilla/5.0
(iPad; CPU OS 7_1_2 like Mac OS X) AppleWebKit/537.51.2 (KHTML, like
Gecko) Version/7.0 Mobile/11D257 Safari/9537.53    fuhongsheng
    _trackPageview        www.baidu.com  10057
1436993753  123.91.75.131   f50cc9c24b1e5cdf06c613702f93ece3
    e5b348e0d13dc2f6967c0759e623e7c4        si.mfniu.com    1680    1050    32

    http://si.mfniu.com/HTML_Content_Cache/stock_indicator/1/000025.h
tml http://si.mfniu.com/choice.aspx Mozilla/4.0 (compatible; MSIE
8.0; Windows NT 6.1; WOW64; Trident/4.0; SLCC2; .NET CLR
2.0.50727; .NET CLR 3.5.30729; .NET CLR 3.0.30729; Media Center PC
6.0)    huatuozhengu    _trackPageview         si.mfniu.com    10119
```

3. 日志字段描述

将一条日志进行切分之后,日志字段执行结果如图 5-2 所示,日志字段描述如表 5-1 所示。

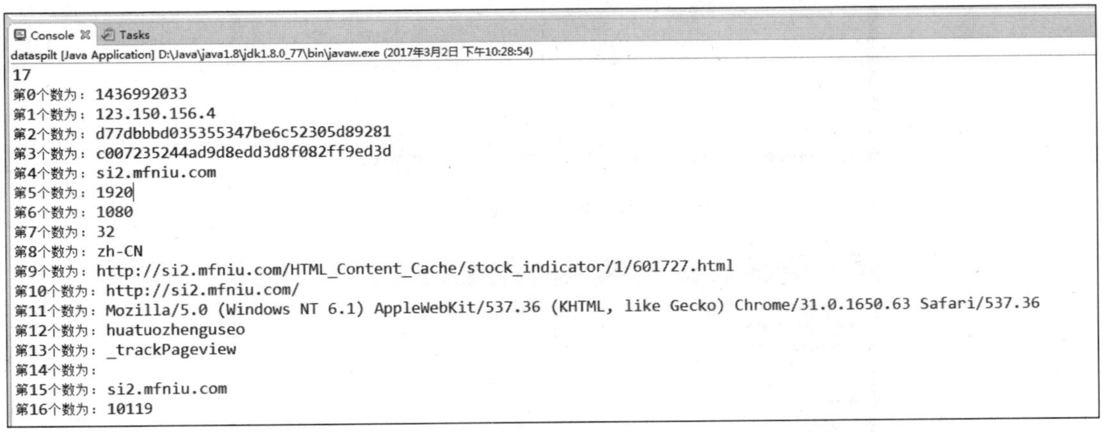

```
Console  Tasks
dataspilt [Java Application] D:\Java\java1.8\jdk1.8.0_77\bin\javaw.exe (2017年3月2日 下午10:28:54)
17
第0个数为: 1436992033
第1个数为: 123.150.156.4
第2个数为: d77dbbbd035355347be6c52305d89281
第3个数为: c007235244ad9d8edd3d8f082ff9ed3d
第4个数为: si2.mfniu.com
第5个数为: 1920
第6个数为: 1080
第7个数为: 32
第8个数为: zh-CN
第9个数为: http://si2.mfniu.com/HTML_Content_Cache/stock_indicator/1/601727.html
第10个数为: http://si2.mfniu.com/
第11个数为: Mozilla/5.0 (Windows NT 6.1) AppleWebKit/537.36 (KHTML, like Gecko) Chrome/31.0.1650.63 Safari/537.36
第12个数为: huatuozhenguseo
第13个数为: _trackPageview
第14个数为:
第15个数为: si2.mfniu.com
第16个数为: 10119
```

图 5-2　日志字段执行结果

表 5-1　日志字段说明

序号	字段	说明
0	timestamp	时间戳
1	user_ip	IP
2	user_tracecode	用户 id

（续表）

序号	字段	说明
3	user_sessionid	会话
4	domain	域名
5	screen_width	屏幕宽
6	screen_height	屏幕高
7	color_dept	颜色深度
8	language	语言
9	url	URL
10	referrer	来源
11	user_agent	浏览器
12	account	追踪账户
13	event	事件（单击，鼠标移动，打开）
14	event_data	事件数据
15	referrer2	来源标记
16	referrer_code	来源编码

4. Eclipse 添加源码

Eclipse 添加源码方法步骤如图 5-3~图 5-5 所示。

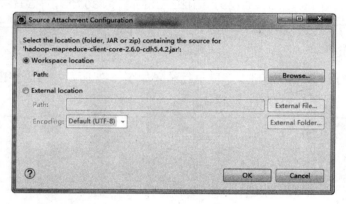

图 5-3　Eclipse 添加源码方法一

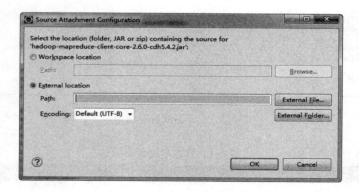

图 5-4　Eclipse 添加源码方法二

下载源码【百度云盘】Soft_BigData，解压放在 D:\Java\hadoop-2.6.0-src 路径下面。

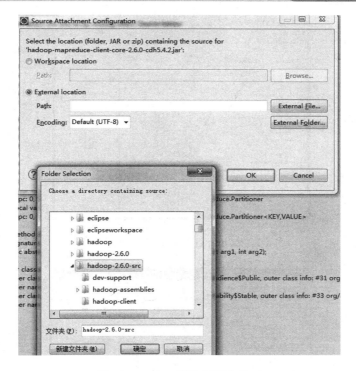

图 5-5　Eclipse 添加源码方法三

5. Eclipse 打包 Runnable Jar

Eclipse 打包 Runnable Jar 步骤如图 5-6~图 5-9 所示。

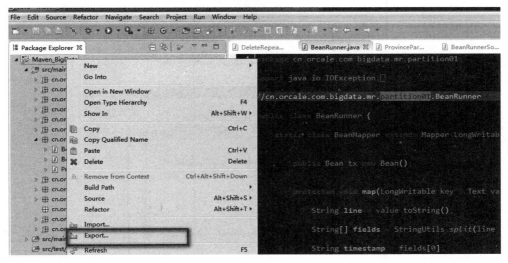

图 5-6　Eclipse 打包 Runnable Jar 步骤一

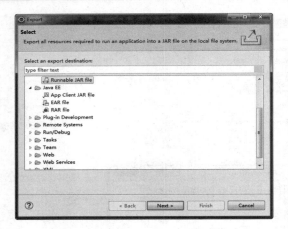

图 5-7　Eclipse 打包 Runnable Jar 步骤二

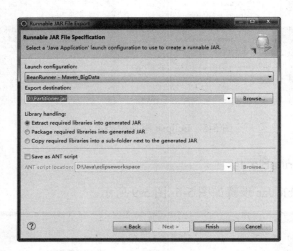

图 5-8　Eclipse 打包 Runnable Jar 步骤三

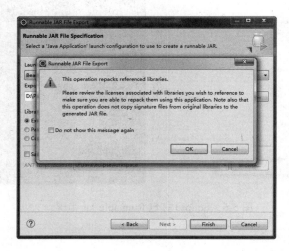

图 5-9　Eclipse 打包 Runnable Jar 步骤四

5.2 查看 YARN 上的任务

执行 MapReduce 过程中，每一个任务都会出现在 YARN 的任务中。YARN 的任务列表如图 5-10 所示。

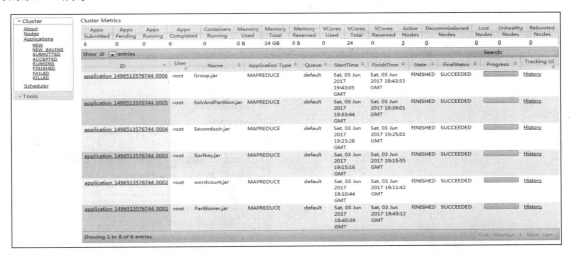

图 5-10　YARN 的任务列表

5.3 加载配置文件

打包之前要将 Hadoop 相关的配置文件加载到 resource 中。Hadoop 配置文件如图 5-11 所示。

图 5-11　Hadoop 配置文件

5.4 MapReduce 实战

1. Java 模仿 WordCount

（1）程序说明。

通过 Java 代码模仿 WordCount。

（2）类的展现，类文件如图 5-12 所示。

（3）结果展现，如图 5-13 所示。

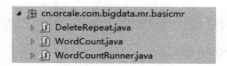

图 5-12 Java 模仿 WordCount 类的展现

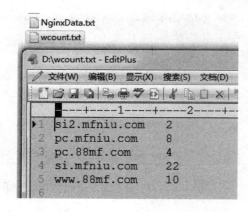

图 5-13 Java 模仿 WordCount 结果展现

（4）代码展现。

```java
package cn.orcale.com.bigdata.mr.basicmr;

import java.io.BufferedReader;
import java.io.File;
import java.io.FileReader;
import java.io.FileWriter;
import java.util.HashMap;
import java.util.Map.Entry;

import org.apache.commons.lang.StringUtils;
import org.apache.hadoop.io.IOUtils;
import org.apache.hadoop.io.Text;


/***
 * 
 * <p>Title:</p>
 * <p>Desc :模仿 wordcount</p>
 * @author yuhui
 * @date 2017年3月2日 下午9:16:24
 */
```

```java
public class WordCount {

    static HashMap<String, Integer> wordsMap = new HashMap<String, Integer>();

    public static void main(String[] args) throws Exception {

        FileReader fileReader = new FileReader("D:\\NginxData.txt");

        BufferedReader br = new BufferedReader(fileReader);

        String line = null;
        while((line=br.readLine())!=null){

            String[] fields = StringUtils.split(line , "\t");

            String domain = fields[4];

            count(domain);

        }

        IOUtils.closeStream(br);

        File file = new File("D:\\wcount.txt");
        if(file.exists()) throw new RuntimeException("运行异常");
        FileWriter fw = new FileWriter(file,true);
        for(Entry<String, Integer> ent : wordsMap.entrySet()){

            fw.write(ent.getKey() + "\t" + ent.getValue());
            fw.write("\r\n");

        }

        IOUtils.closeStream(fw);

    }

    private static void count(String line) {
        String[] words = line.split(" ");
        Integer count = null;

        for(String word:words){

            count = wordsMap.get(word);
            if(count==null){
                wordsMap.put(word, 1);
            }else{
                count ++;
                wordsMap.put(word, count);
```

```
        }
      }
    }
}
```

2. MapReduce 的 WordCount

（1）程序说明

Hadoop 中的 WordCount 简单样例。

（2）类的展现，如图 5-14 所示。

图 5-14　MapReduce 的 WordCount 类的展现

（3）执行展现，如图 5-15 所示。

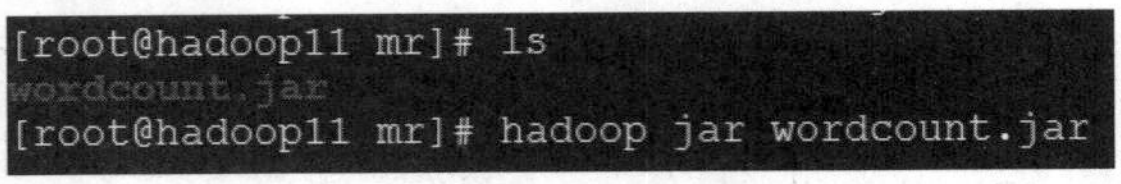

图 5-15　MapReduce 的 WordCount 执行展现

（4）结果展现，如图 5-16 所示。

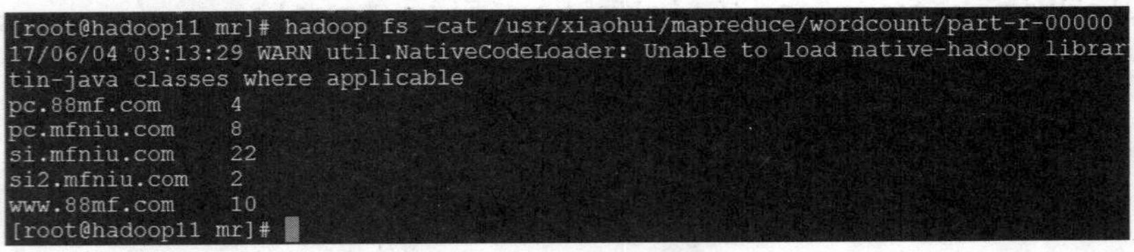

图 5-16　MapReduce 的 WordCount 结果展现

（5）代码展现。

```
package cn.orcale.com.bigdata.mr.basicmr;

import java.io.IOException;

import org.apache.commons.lang.StringUtils;
import org.apache.hadoop.conf.Configuration;
import org.apache.hadoop.fs.FileSystem;
import org.apache.hadoop.fs.Path;
import org.apache.hadoop.io.LongWritable;
import org.apache.hadoop.io.Text;
import org.apache.hadoop.mapreduce.Job;
import org.apache.hadoop.mapreduce.Mapper;
import org.apache.hadoop.mapreduce.Reducer;
import org.apache.hadoop.mapreduce.Mapper.Context;
```

```java
import org.apache.hadoop.mapreduce.lib.input.FileInputFormat;
import org.apache.hadoop.mapreduce.lib.output.FileOutputFormat;

/***
 *
 * <p>
 * Description: 基本的WordCount
 * </p>
 *
 * @author 余辉
 * @date 2017年5月27日下午1:20:15
 * @version 1.0
 */
public class WordCountRunner {

    static class WordCountMapper extends
            Mapper<LongWritable, Text, Text, LongWritable> {

        int count = 0;

        protected void map(LongWritable key, Text value, Context context)
                throws IOException, InterruptedException {

            String line = value.toString();

            String[] fields = StringUtils.split(line, "\t");

            Text domain = new Text(fields[4]);

            context.write(new Text(domain), new LongWritable(1));

        }
    }

    static class WordCountReducer extends
            Reducer<Text, LongWritable, Text, LongWritable> {

        protected void reduce(Text key, Iterable<LongWritable> values,
                Context context) throws IOException,
InterruptedException {

            long counter = 0;

            for (LongWritable value : values) {

                // 累加每一个value
                counter += value.get();
```

```java
        }
            context.write(key, new LongWritable(counter));
        }

    }

    public static void main(String[] args) throws IOException,
        ClassNotFoundException, InterruptedException {

        // 封装任务信息的对象为 Job 对象,所以要先构造一个 Job 对象
        Configuration conf = new Configuration();
        FileSystem fs = FileSystem.get(conf);
        Job job = Job.getInstance(conf);

        // 设置本次 job 作业所在的 jar 包
        job.setJarByClass(WordCountRunner.class);

        // 本次 job 作业使用的 mapper 类是哪个?
        job.setMapperClass(WordCountMapper.class);
        // 本次 job 作业使用的 reducer 类是哪个?
        job.setReducerClass(WordCountReducer.class);

        // 本次 job 作业 mapper 类的输出数据 key 类型
        job.setMapOutputKeyClass(Text.class);
        // 本次 job 作业 mapper 类的输出数据 value 类型
        job.setMapOutputValueClass(LongWritable.class);

        // 本次 job 作业 reducer 类的输出数据 key 类型
        job.setOutputKeyClass(Text.class);
        // 本次 job 作业 reducer 类的输出数据 value 类型
        job.setOutputValueClass(LongWritable.class);

        job.setNumReduceTasks(1);  // 设置 reduce 的个数

        String inputPath = "/usr/xiaohui/mapreduce/data";
        String outputPath = "/usr/xiaohui/mapreduce/wordcount";

        // 判断 output 文件夹是否存在,如果存在则删除
        Path path = new Path(outputPath);// 取第1个表示输出目录参数(第0个参数是输入目录)
        FileSystem fileSystem = path.getFileSystem(conf);// 根据 path 找到这个文件
        if (fileSystem.exists(path)) {
            fileSystem.delete(path, true);// true 的意思是,就算 output 有东西,也一带删除
        }

        // //本次 job 作业要处理的原始数据所在的路径
        FileInputFormat.setInputPaths(job, new Path(inputPath));
```

```
        // //本次job作业产生的结果输出路径
        FileOutputFormat.setOutputPath(job, new Path(outputPath));

        // 提交本次作业
        job.waitForCompletion(true);

    }

}
```

3. MapReduce 的 key 分区

（1）程序说明。

Hadoop 的 Map/Reduce 中支持对 key 进行分区，从而让 Map 出来的数据均匀分布在 Reduce 上，框架自带了一个默认的分区类 Partitioner。

（2）类的展现，如图 5-17 所示。

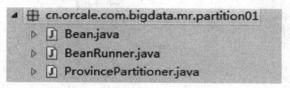

图 5-17　MapReduce 的 key 分区类的展现

（3）执行展现，如图 5-18 所示。

```
[root@hadoop11 mr]# ls
Partitioner.jar  wordcount.jar
[root@hadoop11 mr]# hadoop jar Partitioner.jar
```

图 5-18　MapReduce 的 key 分区执行展现

（4）结果展现，如图 5-19、图 5-20 所示。

Permission	Owner	Group	Size	Replication	Block Size	Name
-rw-r--r--	root	supergroup	0 B	3	128 MB	_SUCCESS
-rw-r--r--	root	supergroup	78 B	3	128 MB	part-r-00000
-rw-r--r--	root	supergroup	832 B	3	128 MB	part-r-00001
-rw-r--r--	root	supergroup	378 B	3	128 MB	part-r-00002
-rw-r--r--	root	supergroup	0 B	3	128 MB	part-r-00003
-rw-r--r--	root	supergroup	454 B	3	128 MB	part-r-00004
-rw-r--r--	root	supergroup	0 B	3	128 MB	part-r-00005

图 5-19　MapReduce 的 key 分区结果展现一

```
[root@hadoop11 mr]# hadoop fs -cat /usr/xiaohui/mapreduce/Partitioner/part-r-00000
17/06/04 02:51:36 WARN util.NativeCodeLoader: Unable to load native-hadoop library
tin-java classes where applicable
si2.mfniu.com    1437992033        123.150.156.4
si2.mfniu.com    1436992033        123.150.156.4
[root@hadoop11 mr]# hadoop fs -cat /usr/xiaohui/mapreduce/Partitioner/part-r-00001
17/06/04 02:51:46 WARN util.NativeCodeLoader: Unable to load native-hadoop library
tin-java classes where applicable
si.mfniu.com    ,1436993124        110.84.190.37
si.mfniu.com     1436993511        1.82.184.71
si.mfniu.com     1436993121        110.84.190.37
si.mfniu.com     1436991332        113.109.196.169
si.mfniu.com     1446913753        123.91.75.131
si.mfniu.com     1436913753        123.91.75.131
si.mfniu.com     1436993503        1.82.184.71
si.mfniu.com     1436993762        123.91.75.131
si.mfniu.com     1438993762        123.91.75.131
si.mfniu.com     1436993364        1.82.184.71
si.mfniu.com     1438991332        113.109.196.169
si.mfniu.com     1438993121        110.84.190.37
si.mfniu.com     1438993124        110.84.190.37
si.mfniu.com     1436991363        113.109.196.169
si.mfniu.com     1438993654        123.91.75.131
si.mfniu.com     1438993503        1.82.184.71
si.mfniu.com     1437993364        1.82.184.71
si.mfniu.com     1436993140        110.84.190.37
si.mfniu.com     1436993654        123.91.75.131
si.mfniu.com     1437991363        113.109.196.169
si.mfniu.com     1437993140        110.84.190.37
si.mfniu.com     1438993511        1.82.184.71
[root@hadoop11 mr]# hadoop fs -cat /usr/xiaohui/mapreduce/Partitioner/part-r-00002
17/06/04 02:51:53 WARN util.NativeCodeLoader: Unable to load native-hadoop library
tin-java classes where applicable
www.88mf.com     1437993038        112.90.231.5
www.88mf.com     1436991538        112.67.214.106
www.88mf.com     1436993038        112.90.231.5
www.88mf.com     1446913101        58.35.139.22
www.88mf.com     1438990555        120.210.170.3
www.88mf.com     1438991345        46.189.73.173
```

图 5-20 MapReduce 的 key 分区结果展现二

（5）代码展现。

```java
package cn.orcale.com.bigdata.mr.partition01;

import java.io.IOException;

import org.apache.commons.lang.StringUtils;
import org.apache.hadoop.conf.Configuration;
import org.apache.hadoop.fs.Path;
import org.apache.hadoop.io.LongWritable;
import org.apache.hadoop.io.Text;
import org.apache.hadoop.mapreduce.Job;
import org.apache.hadoop.mapreduce.Mapper;
import org.apache.hadoop.mapreduce.Reducer;
import org.apache.hadoop.mapreduce.lib.input.FileInputFormat;
import org.apache.hadoop.mapreduce.lib.output.FileOutputFormat;
import org.apache.hadoop.fs.FileSystem;

public class BeanRunner {
```

```java
    static class BeanMapper extends Mapper<LongWritable, Text, Text, Bean> {

        public Bean tx = new Bean();
        protected void map(LongWritable key, Text value, Context context)
                throws IOException, InterruptedException {
            String line = value.toString();
            String[] fields = StringUtils.split(line, "\t");
            String timestamp = fields[0];
            String ip = fields[1];
            Text domain = new Text(fields[4]);
            // 将上下行流量封装到 flowBean 中去
            tx.set(timestamp, ip);
            context.write(domain, tx);

        }

    }

    static class BeanReduce extends Reducer<Text, Bean, Text, Bean> {

        protected void reduce(Text key, Iterable<Bean> values, Context context)
                throws IOException, InterruptedException {

            for (Bean bean : values) {
                context.write(key, bean);
            }

        }

    }

    public static void main(String[] args) throws Exception, IOException,
            InterruptedException {
        Configuration conf = new Configuration();
        Job job = Job.getInstance(conf);

        job.setJarByClass(BeanRunner.class);

        job.setMapperClass(BeanMapper.class);
        job.setReducerClass(BeanReduce.class);

        // 指定自定义的 partitioner 类,替换掉框架默认的 HashPartitioner
```

```java
        job.setPartitionerClass(ProvincePartitioner.class);

        // 指定 reduce task 数量，跟 ProvincePartitioner 的分区数匹配
        job.setNumReduceTasks(6);

        job.setOutputKeyClass(Text.class);
        job.setOutputValueClass(Bean.class);

        String inputPath = "/usr/xiaohui/mapreduce/data";
        String outputPath = "/usr/xiaohui/mapreduce/Partitioner";

        // 判断 output 文件夹是否存在，如果存在则删除
        Path path = new Path(outputPath);// 取第1个表示输出目录参数（第0个参数是输入目录）
        FileSystem fileSystem = path.getFileSystem(conf);// 根据 path 找到这个文件
        if (fileSystem.exists(path)) {
            fileSystem.delete(path, true);// true 的意思是，就算 output 有东西，也一带删除
        }

        // 要处理的数据所在的 path
        // 指定文件夹即可，该文件夹下的所有文件都会被处理
        FileInputFormat.setInputPaths(job, new Path(inputPath));

        // 处理完得到的结果输出的 path
        FileOutputFormat.setOutputPath(job, new Path(outputPath));

        job.waitForCompletion(true);
    }

}
```

```java
package cn.orcale.com.bigdata.mr.partition01;

import java.util.HashMap;

import org.apache.hadoop.io.Text;
import org.apache.hadoop.mapreduce.Partitioner;

public class ProvincePartitioner extends Partitioner<Text, Bean> {

    private static HashMap<String, Integer> provinceMap = new HashMap<String, Integer>();

    static {
        provinceMap.put("si2.mfniu.com", 0);
        provinceMap.put("si.mfniu.com", 1);
        provinceMap.put("www.88mf.com", 2);
    }

    public int getPartition(Text key, Bean value, int numPartitions) {

        Integer province = null;
```

```
            province = provinceMap.get(key.toString());

            if (province == null) {
                province = 4;
            }

            return province;
        }

    }
```

```
package cn.orcale.com.bigdata.mr.partition01;

import java.io.DataInput;
import java.io.DataOutput;
import java.io.IOException;

import org.apache.hadoop.io.Writable;

public class Bean implements Writable {

    private String timestamp;
    private String ip;

    // 因为反序列化时需要反射，必须有一个无参构造函数，在这里显示定义一个
    public Bean() {
    }

    // 定义了一个有参构造函数，会覆盖掉默认的无参构造函数
    public Bean(String timestamp, String ip) {
        this.set(timestamp, ip);
    }

    public void set(String timestamp, String ip) {
        this.timestamp = timestamp;
        this.ip = ip;
    }

    // 序列化方法
    public void write(DataOutput out) throws IOException {
        out.writeUTF(timestamp);
        out.writeUTF(ip);
    }

    // 反序列化方法 ----序列化和反序列化时数据字段的顺序要一致
    public void readFields(DataInput in) throws IOException {
        timestamp = in.readUTF();
        ip = in.readUTF();
    }

    public String toString() {
```

```
        return timestamp + "\t" + ip;
    }

}
```

4. MapReduce 的 key 排序

（1）程序说明。

MapReduce 的 shuffle 过程就是排序过程，key 值会在 shuffle 过程默认按照 tree 的字典排序。

（2）类的展现，如图 5-21 所示。

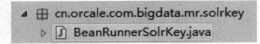

图 5-21　MapReduce 的 key 排序类的展现

（3）执行展现，如图 5-22 所示。

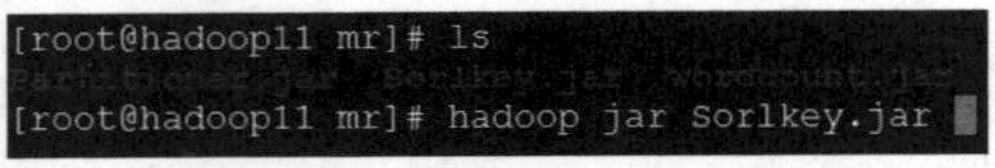

```
[root@hadoop11 mr]# ls
Partitioner.jar  Sorlkey.jar  wordcount.jar
[root@hadoop11 mr]# hadoop jar Sorlkey.jar
```

图 5-22　MapReduce 的 key 排序执行展现

（4）结果展现，如图 5-23 所示。

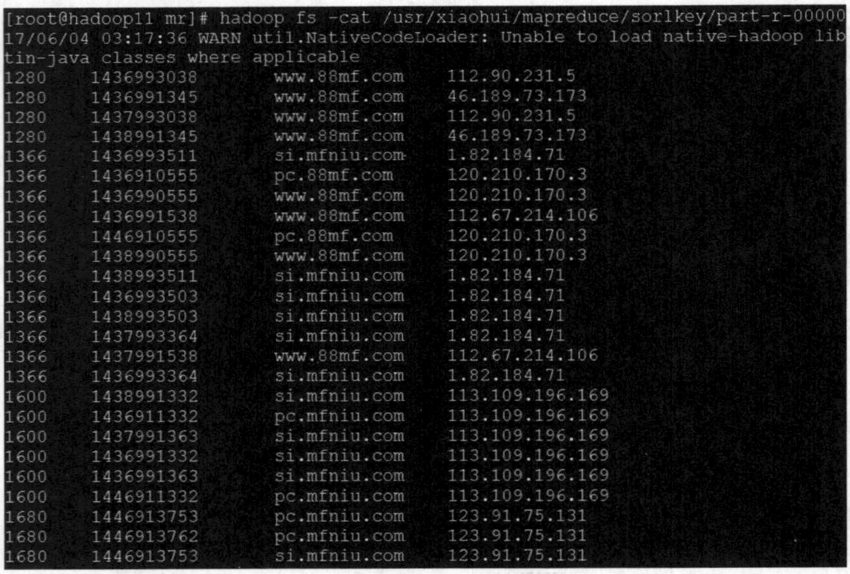

```
[root@hadoop11 mr]# hadoop fs -cat /usr/xiaohui/mapreduce/sorlkey/part-r-00000
17/06/04 03:17:36 WARN util.NativeCodeLoader: Unable to load native-hadoop lib
tin-java classes where applicable
1280    1436993038      www.88mf.com    112.90.231.5
1280    1436991345      www.88mf.com    46.189.73.173
1280    1437993038      www.88mf.com    112.90.231.5
1280    1438991345      www.88mf.com    46.189.73.173
1366    1436993511      si.mfniu.com    1.82.184.71
1366    1436910555      pc.88mf.com     120.210.170.3
1366    1436990555      www.88mf.com    120.210.170.3
1366    1436991538      www.88mf.com    112.67.214.106
1366    1446910555      pc.88mf.com     120.210.170.3
1366    1438990555      www.88mf.com    120.210.170.3
1366    1438993511      si.mfniu.com    1.82.184.71
1366    1436993503      si.mfniu.com    1.82.184.71
1366    1438993503      si.mfniu.com    1.82.184.71
1366    1437993364      si.mfniu.com    1.82.184.71
1366    1437991538      www.88mf.com    112.67.214.106
1366    1436993364      si.mfniu.com    1.82.184.71
1600    1438991332      si.mfniu.com    113.109.196.169
1600    1436911332      pc.mfniu.com    113.109.196.169
1600    1437991363      si.mfniu.com    113.109.196.169
1600    1436991332      si.mfniu.com    113.109.196.169
1600    1436991363      si.mfniu.com    113.109.196.169
1600    1446911332      pc.mfniu.com    113.109.196.169
1680    1446913753      pc.mfniu.com    123.91.75.131
1680    1446913762      pc.mfniu.com    123.91.75.131
1680    1446913753      si.mfniu.com    123.91.75.131
```

图 5-23　MapReduce 的 key 排序的结果展现

(5) 代码展现。

```java
package cn.orcale.com.bigdata.mr.solrkey;

import java.io.IOException;

import org.apache.commons.lang.StringUtils;
import org.apache.hadoop.conf.Configuration;
import org.apache.hadoop.fs.FileSystem;
import org.apache.hadoop.fs.Path;
import org.apache.hadoop.io.LongWritable;
import org.apache.hadoop.io.Text;
import org.apache.hadoop.mapreduce.Job;
import org.apache.hadoop.mapreduce.Mapper;
import org.apache.hadoop.mapreduce.Reducer;
import org.apache.hadoop.mapreduce.lib.input.FileInputFormat;
import org.apache.hadoop.mapreduce.lib.output.FileOutputFormat;

public class BeanRunnerSolrKey {

    static class BeanMapper extends Mapper<LongWritable, Text, Text, Text> {

        protected void map(LongWritable key, Text value, Context context)
                throws IOException, InterruptedException {

            String line = value.toString();

            String[] fields = StringUtils.split(line, "\t");

            String timestamp = fields[0];

            String ip = fields[1];

            String domain = fields[4];

            Text Width = new Text(fields[5]);

            Text tx = new Text(timestamp + " \t " + domain + " \t " + ip);

            context.write(Width, tx);

        }

    }

    static class BeanReduce extends Reducer<Text, Text, Text, Text> {

        protected void reduce(Text key, Iterable<Text> values, Context
```

```java
context)
            throws IOException, InterruptedException {

        for (Text bean : values) {
            context.write(key, bean);
        }

    }

    public static void main(String[] args) throws Exception, IOException,
            InterruptedException {
        Configuration conf = new Configuration();
        Job job = Job.getInstance(conf);

        job.setJarByClass(BeanRunnerSolrKey.class);

        job.setMapperClass(BeanMapper.class);
        job.setReducerClass(BeanReduce.class);

        job.setOutputKeyClass(Text.class);
        job.setOutputValueClass(Text.class);

        String inputPath = "/usr/xiaohui/mapreduce/data";
        String outputPath = "/usr/xiaohui/mapreduce/sorlkey";

        // 判断output文件夹是否存在，如果存在则删除
        Path path = new Path(outputPath);// 取第1个表示输出目录参数（第0个参数是输入目录）
        FileSystem fileSystem = path.getFileSystem(conf);// 根据path找到这个文件
        if (fileSystem.exists(path)) {
            fileSystem.delete(path, true);// true的意思是，就算output有东西，也一带删除
        }

        // 要处理的数据所在的path
        // 指定文件夹即可，该文件夹下的所有文件都会被处理
        FileInputFormat.setInputPaths(job, new Path(inputPath));

        // 处理完得到结果输出的path
        FileOutputFormat.setOutputPath(job, new Path(outputPath));

        job.waitForCompletion(true);
    }

}
```

5. MapReduce 的第二次排序

（1）程序说明。

MapReduce 支持自定义排序，Map 端将 key 值自定义一个 Bean 类。在 Bean 中实现 WritableComparable 类，重写 compareTo 方法。

（2）类的展现，如图 5-24 所示。

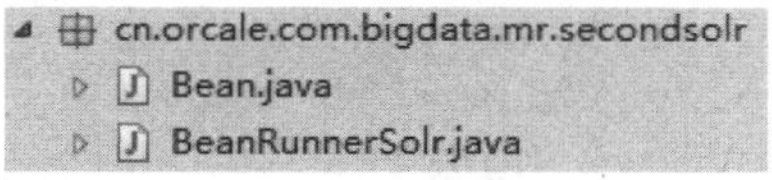

图 5-24　MapReduce 第二次排序类的展现

（3）执行展现，如图 5-25 所示。

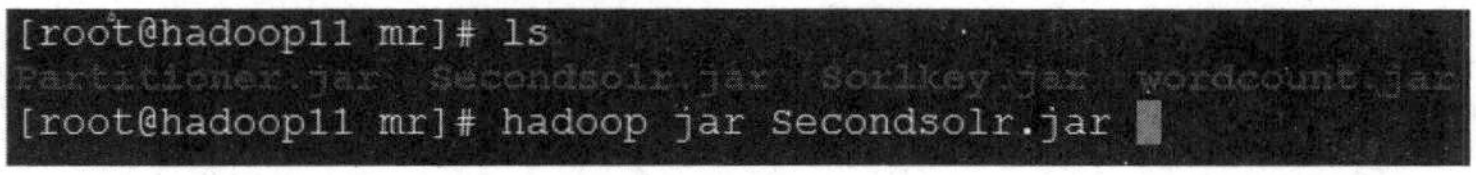

```
[root@hadoop11 mr]# ls
Partitioner.jar  Secondsolr.jar  Sorlkey.jar  wordcount.jar
[root@hadoop11 mr]# hadoop jar Secondsolr.jar
```

图 5-25　MapReduce 第二次排序执行展现

（4）结果展现，如图 5-26 所示。

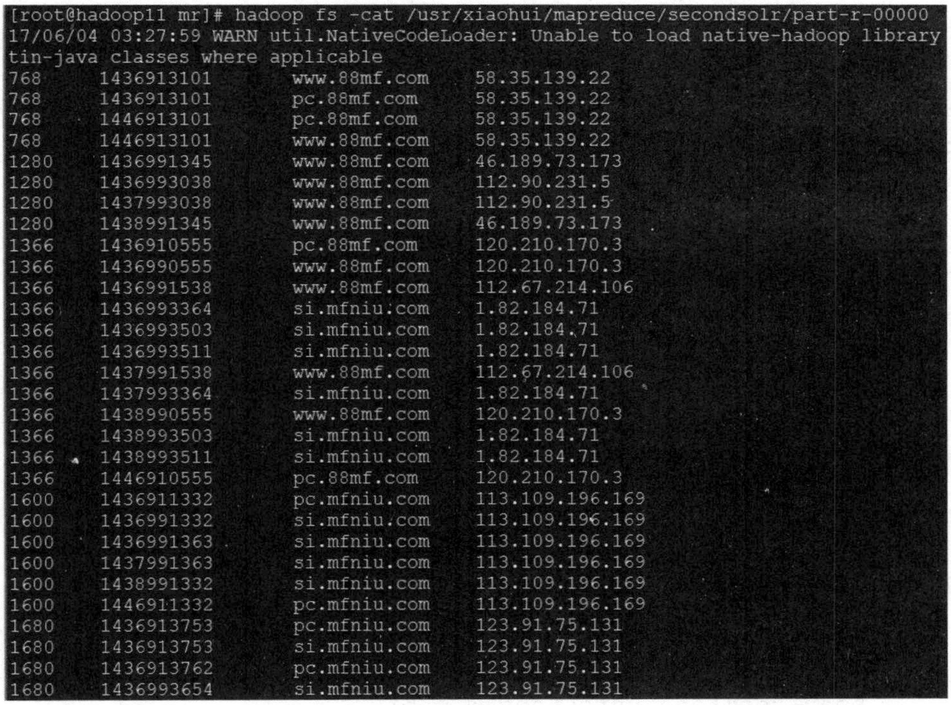

```
[root@hadoop11 mr]# hadoop fs -cat /usr/xiaohui/mapreduce/secondsolr/part-r-00000
17/06/04 03:27:59 WARN util.NativeCodeLoader: Unable to load native-hadoop library
tin-java classes where applicable
768     1436913101      www.88mf.com    58.35.139.22
768     1436913101      pc.88mf.com     58.35.139.22
768     1446913101      pc.88mf.com     58.35.139.22
768     1446913101      www.88mf.com    58.35.139.22
1280    1436991345      www.88mf.com    46.189.73.173
1280    1436993038      www.88mf.com    112.90.231.5
1280    1437993038      www.88mf.com    112.90.231.5
1280    1438991345      www.88mf.com    46.189.73.173
1366    1436910555      pc.88mf.com     120.210.170.3
1366    1436990555      www.88mf.com    120.210.170.3
1366    1436991538      www.88mf.com    112.67.214.106
1366    1436993364      si.mfniu.com    1.82.184.71
1366    1436993503      si.mfniu.com    1.82.184.71
1366    1436993511      si.mfniu.com    1.82.184.71
1366    1437991538      www.88mf.com    112.67.214.106
1366    1437993364      si.mfniu.com    1.82.184.71
1366    1438990555      www.88mf.com    120.210.170.3
1366    1438993503      si.mfniu.com    1.82.184.71
1366    1438993511      si.mfniu.com    1.82.184.71
1366    1446910555      pc.88mf.com     120.210.170.3
1600    1436911332      pc.mfniu.com    113.109.196.169
1600    1436991332      si.mfniu.com    113.109.196.169
1600    1436991363      si.mfniu.com    113.109.196.169
1600    1437991363      si.mfniu.com    113.109.196.169
1600    1438991332      si.mfniu.com    113.109.196.169
1600    1446911332      pc.mfniu.com    113.109.196.169
1680    1436913753      pc.mfniu.com    123.91.75.131
1680    1436913753      si.mfniu.com    123.91.75.131
1680    1436913762      pc.mfniu.com    123.91.75.131
1680    1436993654      si.mfniu.com    123.91.75.131
```

图 5-26　MapReduce 第二次排序结果展现

(5) 代码展现。

```java
package cn.orcale.com.bigdata.mr.secondsolr;

import java.io.IOException;

import org.apache.commons.lang.StringUtils;
import org.apache.hadoop.conf.Configuration;
import org.apache.hadoop.fs.FileSystem;
import org.apache.hadoop.fs.Path;
import org.apache.hadoop.io.LongWritable;
import org.apache.hadoop.io.Text;
import org.apache.hadoop.mapreduce.Job;
import org.apache.hadoop.mapreduce.Mapper;
import org.apache.hadoop.mapreduce.Reducer;
import org.apache.hadoop.mapreduce.lib.input.FileInputFormat;
import org.apache.hadoop.mapreduce.lib.output.FileOutputFormat;

public class BeanRunnerSolr {

    static class BeanMapper extends Mapper<LongWritable, Text, Bean, Text> {

        public Bean bean = new Bean();

        protected void map(LongWritable key, Text value, Context context)
                throws IOException, InterruptedException {

            String line = value.toString();

            String[] fields = StringUtils.split(line, "\t");

            String timestamp = fields[0];

            String ip = fields[1];

            String domain = fields[4];

            String Width = fields[5];

            Text tx = new Text(timestamp + " \t " + domain + " \t " + ip);

            bean.set(Long.parseLong(Width), Long.parseLong(timestamp));

            context.write(bean, tx);

        }
```

```java
    }

    static class BeanReduce extends Reducer<Bean, Text, LongWritable, Text> {

        protected void reduce(Bean key, Iterable<Text> values, Context context)
                throws IOException, InterruptedException {

            for (Text bean : values) {
                context.write(new LongWritable(key.getWidth()), bean);
            }

        }

    }

    public static void main(String[] args) throws Exception, IOException,
            InterruptedException {
        Configuration conf = new Configuration();
        Job job = Job.getInstance(conf);

        job.setJarByClass(BeanRunnerSolr.class);

        job.setMapperClass(BeanMapper.class);
        job.setReducerClass(BeanReduce.class);

        job.setOutputKeyClass(Bean.class);
        job.setOutputValueClass(Text.class);

        String inputPath = "/usr/xiaohui/mapreduce/data";
        String outputPath = "/usr/xiaohui/mapreduce/secondsolr";

        // 判断output文件夹是否存在，如果存在则删除
        Path path = new Path(outputPath);// 取第1个表示输出目录参数（第0个参数是输入目录）
        FileSystem fileSystem = path.getFileSystem(conf);// 根据path找到这个文件
        if (fileSystem.exists(path)) {
            fileSystem.delete(path, true);// true的意思是，就算output有东西，也一带删除
        }

        // //本次job作业要处理的原始数据所在的路径
        FileInputFormat.setInputPaths(job, new Path(inputPath));
        // //本次job作业产生的结果输出路径
        FileOutputFormat.setOutputPath(job, new Path(outputPath));

        job.waitForCompletion(true);
```

```
    }
}
```

```java
package cn.orcale.com.bigdata.mr.secondsolr;

import java.io.DataInput;
import java.io.DataOutput;
import java.io.IOException;

import org.apache.hadoop.io.WritableComparable;

public class Bean implements WritableComparable<Bean> {

    private long Width;
    private long timestamp;

    // 因为反序列化时需要反射，必须有一个无参构造函数，在这里显示定义一个
    public Bean() {
    }

    // 定义了一个有参构造函数，会覆盖掉默认的无参构造函数
    public Bean(long Width, long timestamp) {
        this.set(Width, timestamp);
    }

    public void set(long Width, long timestamp) {
        this.Width = Width;
        this.timestamp = timestamp;
    }

    public long getWidth() {
        return Width;
    }

    // 序列化方法
    public void write(DataOutput out) throws IOException {
        out.writeLong(Width);
        out.writeLong(timestamp);

    }

    // 反序列化方法 ----序列化和反序列化时数据字段的顺序要一致
    public void readFields(DataInput in) throws IOException {
        Width = in.readLong();
        timestamp = in.readLong();
    }

    public int compareTo(Bean Key) {
        long min = Width - Key.Width;
```

```
            if (min != 0) {
                // 说明第一列不相等，则返回两数之间小的数
                return (int) min;
            } else {
                return (int) (timestamp - Key.timestamp);
            }
        }
    }
}
```

6. MapReduce 的排序和分区

（1）程序说明。

本节中按照 domain 分区，按照 width 排序。

（2）类的展现，如图 5-27 所示。

图 5-27　MapReduce 排序和分区类的展现

（3）执行展现，如图 5-28 所示。

图 5-28　MapReduce 排序和分区执行展现

（4）结果展现，如图 5-29、图 5-30 所示。

/usr/xiaohui/mapreduce/solrandpartition

Permission	Owner	Group	Size	Replication	Block Size	Name
-rw-r--r--	root	supergroup	0 B	3	128 MB	_SUCCESS
-rw-r--r--	root	supergroup	92 B	3	128 MB	part-r-00000
-rw-r--r--	root	supergroup	986 B	3	128 MB	part-r-00001
-rw-r--r--	root	supergroup	446 B	3	128 MB	part-r-00002
-rw-r--r--	root	supergroup	0 B	3	128 MB	part-r-00003
-rw-r--r--	root	supergroup	536 B	3	128 MB	part-r-00004
-rw-r--r--	root	supergroup	0 B	3	128 MB	part-r-00005

图 5-29　MapReduce 排序和分区结果展现一

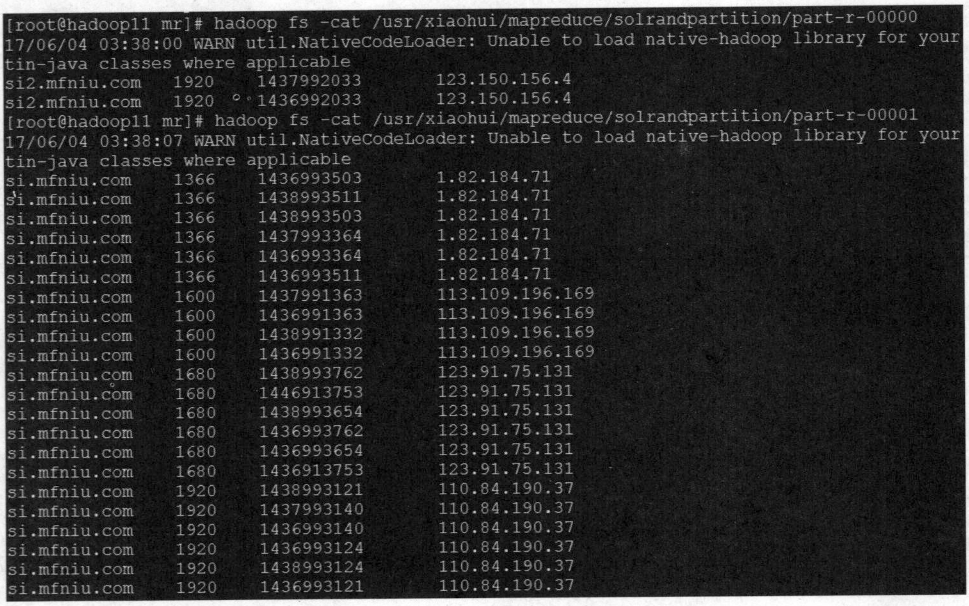

```
[root@hadoop11 mr]# hadoop fs -cat /usr/xiaohui/mapreduce/solrandpartition/part-r-00000
17/06/04 03:38:00 WARN util.NativeCodeLoader: Unable to load native-hadoop library for your
tin-java classes where applicable
si2.mfniu.com    1920    1437992033        123.150.156.4
si2.mfniu.com    1920    1436992033        123.150.156.4
[root@hadoop11 mr]# hadoop fs -cat /usr/xiaohui/mapreduce/solrandpartition/part-r-00001
17/06/04 03:38:07 WARN util.NativeCodeLoader: Unable to load native-hadoop library for your
tin-java classes where applicable
si.mfniu.com     1366    1436993503        1.82.184.71
si.mfniu.com     1366    1438993511        1.82.184.71
si.mfniu.com     1366    1438993503        1.82.184.71
si.mfniu.com     1366    1437993364        1.82.184.71
si.mfniu.com     1366    1436993364        1.82.184.71
si.mfniu.com     1366    1436993511        1.82.184.71
si.mfniu.com     1600    1437991363        113.109.196.169
si.mfniu.com     1600    1436991363        113.109.196.169
si.mfniu.com     1600    1438991332        113.109.196.169
si.mfniu.com     1600    1436991332        113.109.196.169
si.mfniu.com     1680    1438993762        123.91.75.131
si.mfniu.com     1680    1446913753        123.91.75.131
si.mfniu.com     1680    1438993654        123.91.75.131
si.mfniu.com     1680    1436993762        123.91.75.131
si.mfniu.com     1680    1436993654        123.91.75.131
si.mfniu.com     1680    1436913753        123.91.75.131
si.mfniu.com     1920    1438993121        110.84.190.37
si.mfniu.com     1920    1437993140        110.84.190.37
si.mfniu.com     1920    1436993140        110.84.190.37
si.mfniu.com     1920    1436993124        110.84.190.37
si.mfniu.com     1920    1438993124        110.84.190.37
si.mfniu.com     1920    1436993121        110.84.190.37
```

图 5-30　MapReduce 排序和分区结果展现二

（5）代码展现。

```java
package cn.orcale.com.bigdata.mr.partitionsolr;

import java.io.IOException;

import org.apache.commons.lang.StringUtils;
import org.apache.hadoop.conf.Configuration;
import org.apache.hadoop.fs.FileSystem;
import org.apache.hadoop.fs.Path;
import org.apache.hadoop.io.LongWritable;
import org.apache.hadoop.io.Text;
import org.apache.hadoop.mapreduce.Job;
import org.apache.hadoop.mapreduce.Mapper;
import org.apache.hadoop.mapreduce.Reducer;
import org.apache.hadoop.mapreduce.lib.input.FileInputFormat;
import org.apache.hadoop.mapreduce.lib.output.FileOutputFormat;

public class BeanRunnerPartitionsolr {

    static class BeanMapper extends Mapper<LongWritable, Text, Bean, Text> {

        public Bean bean = new Bean();

        protected void map(LongWritable key, Text value, Context context)
                throws IOException, InterruptedException {
```

```java
            String line = value.toString();
            String[] fields = StringUtils.split(line, "\t");
            String timestamp = fields[0];
            String ip = fields[1];
            String domain = fields[4];
            String Width = fields[5];
            Text tx = new Text(timestamp + " \t " + ip);
            bean.set(domain, Long.parseLong(Width));
            context.write(bean, tx);
        }
    }

    static class BeanReduce extends Reducer<Bean, Text, Bean, Text> {
        protected void reduce(Bean key, Iterable<Text> values, Context context)
                throws IOException, InterruptedException {
            for (Text bean : values) {
                context.write(key, bean);
            }
        }
    }

    public static void main(String[] args) throws Exception, IOException,
            InterruptedException {
        Configuration conf = new Configuration();
        Job job = Job.getInstance(conf);

        job.setJarByClass(BeanRunnerPartitionsolr.class);

        job.setMapperClass(BeanMapper.class);
        job.setReducerClass(BeanReduce.class);

        // 指定自定义的partitioner类,替换掉框架默认的HashPartitioner
        job.setPartitionerClass(ProvincePartitioner.class);
```

```java
        // 指定 reduce task 数量，跟 ProvincePartitioner 的分区数匹配
        job.setNumReduceTasks(6);

        job.setOutputKeyClass(Bean.class);
        job.setOutputValueClass(Text.class);

        String inputPath = "/usr/xiaohui/mapreduce/data";
        String outputPath = "/usr/xiaohui/mapreduce/solrandpartition";

        // 判断 output 文件夹是否存在，如果存在则删除
        Path path = new Path(outputPath);// 取第1个表示输出目录参数（第0个参数是输入目录）
        FileSystem fileSystem = path.getFileSystem(conf);// 根据path找到这个文件
        if (fileSystem.exists(path)) {
            fileSystem.delete(path, true);// true 的意思是，就算 output 有东西，也一带删除
        }

        // //本次 job 作业要处理的原始数据所在的路径
        FileInputFormat.setInputPaths(job, new Path(inputPath));
        // //本次 job 作业产生的结果输出路径
        FileOutputFormat.setOutputPath(job, new Path(outputPath));

        job.waitForCompletion(true);
    }

}
```

```java
package cn.orcale.com.bigdata.mr.partitionsolr;

import java.util.HashMap;

import org.apache.hadoop.io.Text;
import org.apache.hadoop.mapreduce.Partitioner;

public class ProvincePartitioner extends Partitioner<Bean, Text> {

    private static HashMap<String, Integer> provinceMap = new HashMap<String, Integer>();

    static {

        provinceMap.put("si2.mfniu.com", 0);
        provinceMap.put("si.mfniu.com", 1);
        provinceMap.put("www.88mf.com", 2);
    }

    public int getPartition(Bean key, Text value, int numPartitions) {
```

```
            Integer province = null;
            province = provinceMap.get(key.getDomain().toString());

            if (province == null) {
                province = 4;
            }

            return province;
        }

}
```

```
package cn.orcale.com.bigdata.mr.partitionsolr;

import java.io.DataInput;
import java.io.DataOutput;
import java.io.IOException;

import org.apache.hadoop.io.WritableComparable;

public class Bean implements WritableComparable<Bean> {

    private String domain;
    private long Width;

    // 因为反序列化时需要反射，必须有一个无参构造函数，在这里显示定义一个
    public Bean() {
    }

    // 定义了一个有参构造函数，会覆盖掉默认的无参构造函数
    public Bean(String domain, long Width) {
        this.set(domain, Width);
    }

    public void set(String domain, long Width) {
        this.domain = domain;
        this.Width = Width;
    }

    public String getDomain() {
        return domain;
    }

    public long getWidth() {
        return Width;
    }

    // 序列化方法
    public void write(DataOutput out) throws IOException {
```

```java
    out.writeUTF(domain);
    out.writeLong(Width);
}

// 反序列化方法 ----序列化和反序列化时数据字段的顺序要一致
public void readFields(DataInput in) throws IOException {
    domain = in.readUTF();
    Width = in.readLong();
}

public String toString() {

    return domain + "\t" + Width;
}

public int compareTo(Bean Key) {

    return (int) (Width - Key.Width);

}
}
```

7. MapReduce 的分组

（1）程序说明。

通过 MapReduce 中的 key 进行分组，在 Reduce 端找出 Domain 中最大屏幕宽。

（2）类的展现，如图 5-31 所示。

图 5-31　MapReduce 分组类的展现

（3）执行展现，如图 5-32 所示。

图 5-32　MapReduce 分组执行展现

（4）结果展现，如图 5-33 所示。

图 5-33　MapReduce 分组结果展现

（5）代码展现。

```java
package cn.orcale.com.bigdata.mr.groupdefault;

import java.io.IOException;

import org.apache.commons.lang.StringUtils;
import org.apache.hadoop.conf.Configuration;
import org.apache.hadoop.fs.FileSystem;
import org.apache.hadoop.fs.Path;
import org.apache.hadoop.io.LongWritable;
import org.apache.hadoop.io.Text;
import org.apache.hadoop.mapreduce.Job;
import org.apache.hadoop.mapreduce.Mapper;
import org.apache.hadoop.mapreduce.Reducer;
import org.apache.hadoop.mapreduce.lib.input.FileInputFormat;
import org.apache.hadoop.mapreduce.lib.output.FileOutputFormat;

public class BeanRunnerGroup {

    static class BeanMapper extends Mapper<LongWritable, Text, Text, Text> {

        protected void map(LongWritable key, Text value, Context context)
                throws IOException, InterruptedException {

            String line = value.toString();

            String[] fields = StringUtils.split(line, "\t");

            Text domain = new Text(fields[4]);

            Text Width = new Text(fields[5]);

            context.write(domain, Width);

        }

    }

    static class BeanReduce extends Reducer<Text, Text, Text, Text> {

        protected void reduce(Text key, Iterable<Text> values, Context context)
                throws IOException, InterruptedException {

            String temp = "0";

            for (Text bean : values) {
```

```java
                    if (bean.toString().compareTo(temp) > 0) {
                        temp = bean.toString();
                    }
                }
                context.write(key, new Text(temp));
            }
        }

        public static void main(String[] args) throws Exception,
IOException,
                InterruptedException {
            Configuration conf = new Configuration();
            Job job = Job.getInstance(conf);

            job.setJarByClass(BeanRunnerGroup.class);

            job.setMapperClass(BeanMapper.class);
            job.setReducerClass(BeanReduce.class);

            job.setOutputKeyClass(Text.class);
            job.setOutputValueClass(Text.class);

            String inputPath = "/usr/xiaohui/mapreduce/data";
            String outputPath = "/usr/xiaohui/mapreduce/Group";

            // 判断 output 文件夹是否存在，如果存在则删除
            Path path = new Path(outputPath);// 取第1个表示输出目录参数（第0个参数是输入目录）
            FileSystem fileSystem = path.getFileSystem(conf);// 根据 path 找到这个文件
            if (fileSystem.exists(path)) {
                fileSystem.delete(path, true);// true 的意思是，就算 output 有东西，也一带删除
            }

            // 要处理的数据所在的 path
            // 指定文件夹即可，该文件夹下的所有文件都会被处理
            FileInputFormat.setInputPaths(job, new Path(inputPath));

            // 处理完得到的结果输出的 path
            FileOutputFormat.setOutputPath(job, new Path(outputPath));

            job.waitForCompletion(true);
        }
}
```

5.5 小结

　　MapReduce 是一种编程模型，用于大规模数据集（大于 1TB）的并行运算。本章讲解 MapReduce 组件，首先准备日志数据且讲解日志字段，为 MapReduce 实战做好前期工作。其次图解 Eclipse 添加源码和 Eclipse 打包 Runnable Jar，为 MapReduce 实战做好铺垫。最后 MapReduce 实战分别从程序说明、类的展示、执行展示、结果展示、代码展示 5 个维度详细描述 MapReduce 的 WordCount、key 分区、key 排序、第二次排序、排序和分区、分组实战。

第 6 章

◀ HBase实战 ▶

6.1 HBase 简介及架构

HBase 是一种构建在 HDFS 之上的分布式、面向列的存储系统。在需要实时读写、随机访问超大规模数据集时，可以使用 HBase。

尽管已经有许多数据存储和访问的策略和实现方法，但事实上大多数解决方案，特别是一些关系类型的，在构建时并没有考虑超大规模和分布式的特点。许多商家通过复制和分区的方法来扩充数据库，使其突破单个节点的界限，但这些功能通常都是事后增加的，安装和维护都很复杂。同时，也会影响 RDBMS 的特定功能，例如联接、复杂的查询、触发器、视图和外键约束，这些操作在大型的 RDBMS 上的代价相当高，甚至根本无法实现。

HBase 从另一个角度处理伸缩性问题。它通过线性方式从下到上增加节点来进行扩展。HBase 不是关系型数据库，也不支持 SQL，但是它有自己的特长，这是 RDBMS 不能处理的，HBase 巧妙地将大而稀疏的表放在商用的服务器集群上。

HBase 是 Google Bigtable 的开源实现，与 Google Bigtable 利用 GFS 作为其文件存储系统类似，HBase 利用 Hadoop HDFS 作为其文件存储系统；Google 运行 MapReduce 来处理 Bigtable 中的海量数据，HBase 同样利用 Hadoop MapReduce 来处理 HBase 中的海量数据；Google Bigtable 利用 Chubby 作为协同服务，HBase 利用 ZooKeeper 作为对应。

1. HBase 的特点

- 大：一个表可以有上亿行，上百万列。
- 面向列：面向列表（簇）的存储和权限控制，列（簇）独立检索。
- 稀疏：对于为空（NULL）的列，并不占用存储空间，因此，表可以设计得非常稀疏。
- 无模式：每一行都有一个可以排序的主键和任意多的列，列可以根据需要动态增加，同一张表中不同的行可以有截然不同的列。
- 数据多版本：每个单元中的数据可以有多个版本，默认情况下，版本号自动分配，版本号就是单元格插入时的时间戳。
- 数据类型单一：HBase 中的数据都是字符串，没有类型。

2. HBase 系统架构

HBase 系统架构如图 6-1 所示。

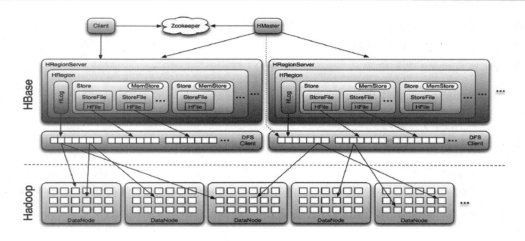

图 6-1　HBase 系统架构

从 HBase 的架构图上可以看出，HBase 中的组件包括 Client、ZooKeeper、HMaster、HRegionServer、HRegion、Store、MemStore、StoreFile、HFile、HLog 等，接下来介绍部分作用。

HBase 中的每张表都通过行键按照一定的范围被分割成多个子表（HRegion），默认一个 HRegion 超过 256MB 就要被分割成两个，这个过程由 HRegionServer 管理，而 HRegion 的分配由 HMaster 管理。

（1）Client

包含访问 HBase 的接口，Client 维护着一些 cache 来加快对 HBase 的访问，比如 Region 的位置信息。

（2）ZooKeeper 的作用

HBase 依赖 ZooKeeper，默认情况下 HBase 管理 ZooKeeper 实例（启动或关闭 ZooKeeper），Master 与 RegionServers 启动时会向 ZooKeeper 注册。ZooKeeper 在 HBase 中功能如图 6-2 所示。

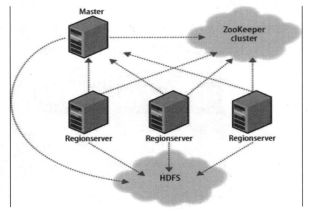

图 6-2　ZooKeeper 在 HBase 中功能

ZooKeeper 的作用：

- 保证任何时候，集群中只有一个 Master。
- 存储所有 Region 的寻址入口。
- 实时监控 Region Server 的上线和下线信息，并实时通知给 Master。
- 存储 HBase 的 schema 和 table 元数据。

（3）HMaster 的作用

- 为 Region Server 分配 Region。
- 负责 Region Server 的负载均衡。

- 发现失效的 Region Server 并重新分配其上的 Region。
- HDFS 上的垃圾文件回收。
- 处理 schema 更新请求。

（4）HRegionServer 的作用

- 维护 Master 分配给他的 Region，处理对这些 Region 的 IO 请求。
- 负责切分正在运行过程中变得过大的 Region 。

注意：Client 访问 HBase 上的数据时不需要 Master 的参与，因为数据寻址访问 ZooKeeper 和 Region Server，而数据读写访问 Region Server。Master 仅仅维护 table 和 Region 的元数据信息，而 table 的元数据信息保存在 ZooKeeper 上，因此 Master 负载很低。

（5）HRegion

table 在行的方向上分隔为多个 Region。Region 是 HBase 中分布式存储和负载均衡的最小单元，即不同的 Region 可以分别在不同的 Region Server 上，但同一个 Region 是不会拆分到多个 Server 上。

Region 按大小分隔，每个表一般是只有一个 Region 。随着数据不断插入表，Region 不断增大，当 Region 的某个列族达到一个阈值（默认 256M）时就会分成两个新的 Region 。

3. HBase 逻辑模型

HBase 以表的形式存储数据。表由行和列组成。列划分为若干个列族（Column family），HBase 逻辑模型如图 6-3 所示。

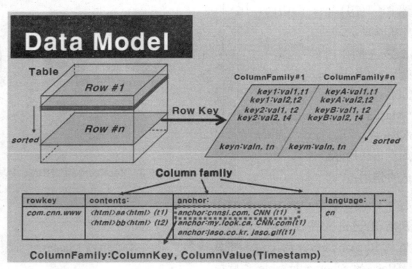

图 6-3　HBase 逻辑模型

HBase 的逻辑数据模型如图 6-4 所示。

Row Key	Time Stamp	Column "contents:"	Column "anchor:"		Column "mime:"
"com.cnn.www"	t9		"anchor:cnnsi.com"	"CNN"	
	t8		"anchor:my.look.ca"	"CNN.com"	
	t6	"<html>..."			"text/html"
	t5	"<html>..."			
	t3	"<html>..."			

图 6-4 HBase 的逻辑数据模型

HBase 的物理数据模型，即实际存储的数据模型如图 6-5 所示。

Row Key	Time Stamp	Column "contents:"
"com.cnn.www"	t6	"<html>..."
	t5	"<html>..."
	t3	"<html>..."

Row Key	Time Stamp	Column "anchor:"	
"com.cnn.www"	t9	"anchor:cnnsi.com"	"CNN"
	t8	"anchor:my.look.ca"	"CNN.com"

Row Key	Time Stamp	Column "mime:"
"com.cnn.www"	t6	"text/html"

图 6-5 HBase 的物理数据模型

逻辑数据模型中空白 Cell 在物理上是不存储的，因为根本没有必要存储，因此若一个请求为要获取 t8 时间的 contents:html，它的结果就是空。与这个相似，若请求为获取 t9 时间的 anchor:my.look.ca，结果也是空。但是，如果不指明时间，将会返回最新时间的行，每个最新的都会返回。

（1）Row Key

与 NoSQL 数据库一样，Row Key 是用来检索记录的主键。访问 HBase table 中的行，只有三种方式：

- 通过单个 Row Key 访问。
- 通过 Row Key 的 range 全表扫描。
- Row Key 可以是任意字符串（最大长度是 64KB，实际应用中长度一般为 10～100bytes），在 HBase 内部，Row Key 保存为字节数组。

在存储时，数据按照 Row Key 的字典序（byte order）排序存储。设计 Key 时，要充分考滤排序存储这个特性，将经常一起读取的行存储到一起（位置相关性）。

注意：字典序对 int 排序的结果是 1，10,100,11,12,13,14,15，16,17,18,19，20,21，…，9，91,92,93,94,95,96,97,98,99。要保存整形的自然序，Row Key 必须用 0 进行左填充。

行的一次读写是原子操作（不论一次读写多少列）。这个设计决策能够使用户很容易地理解程序在对同一个行进行并发更新操作时的行为。

```
0-9（对应数值48-59）;
A-Z（对应数值65-90）;
a-z（对应数值97-122）;
```

（2）列族

HBase 表中的每个列都归属于某个列族。列族是表的 Schema 的一部分（而列不是），必须在使用表之前定义。列名都以列族作为前缀，例如 courses:history、courses:math 都属于 courses 这个列族。

访问控制、磁盘和内存的使用统计都是在列族层面进行的。在实际应用中，列族上的控制权限能帮助我们管理不同类型的应用，例如，允许一些应用可以添加新的基本数据、一些应用可以读取基本数据并创建继承的列族、一些应用则只允许浏览数据（甚至可能因为隐私的原因不能浏览所有数据）。

（3）时间戳

HBase 中通过 Row 和 Columns 确定的一个存储单元称为 Cell。每个 Cell 都保存着同一份数据的多个版本。版本通过时间戳来索引，时间戳的类型是 64 位整型。时间戳可以由 HBase（在数据写入时自动）赋值，此时时间戳是精确到毫秒的当前系统时间。时间戳也可以由客户显式赋值。如果应用程序要避免数据版本冲突，就必须自己生成具有唯一性的时间戳。每个 Cell 中，不同版本的数据按照时间倒序排序，即最新的数据排在最前面。

为了避免数据存在过多版本造成的管理（包括存储和索引）负担，HBase 提供了两种数据版本回收方式。一是保存数据的最后 n 个版本，二是保存最近一段时间内的版本（比如最近 7 天）。用户可以针对每个列族进行设置。

（4）Cell

Cell 是由 {row key, column(=< family> + < label>), version} 唯一确定的单元。Cell 中的数据是没有类型的，全部是字节码形式存储。

4. HBase 物理存储

Table 在行的方向上分割为多个 HRegion，每个 HRegion 分散在不同的 Region Server 中。HBase 物理存储如图 6-6 所示。

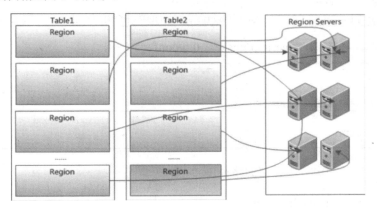

图 6-6　HBase 物理存储

每个 HRegion 由多个 Store 构成，每个 Store 由一个 memStore 和 0 或多个 StoreFile 组成，每个 Store 保存一个 Columns Family。HRegion 和 Store 关系如图 6-7 所示。

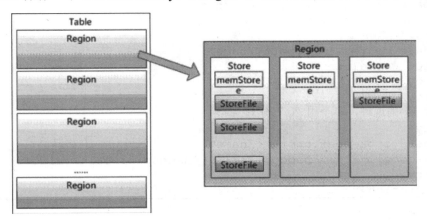

图 6-7　HRegion 和 Store 关系

6.2　HBase 安装

（1）上传 HBase 安装包 hbase-1.0.0-bin.tar.gz。
（2）解压 hbase-1.0.0-bin.tar.gz。
（3）配置 HBase 集群，要修改 3 个文件（首先 zk 集群已经安装好了 HMASTER REGIONSERVER）。注意：要把 Hadoop 的 hdfs-site.xml 和 core-site.xml 放到 hbase/conf 下。

vi /etc/profile

```
export JAVA_HOME=/usr/app/jdk1.7
export HADOOP_HOME=/usr/app/hadoop-2.6.0
export HBASE_HOME=/usr/app/hbase-1.0.0
export PATH=$PATH:$JAVA_HOME/bin:$HADOOP_HOME/bin:$HBASE_HOME/bin
```

（4）修改 hbase-env.sh，/usr/app/hbase-1.0.0/conf。

```
export JAVA_HOME=/usr/app/jdk1.7
//告诉HBase使用外部的zk
export HBASE_MANAGES_ZK=false
```

vim hbase-site.xml

```
<configuration>
<!-- 指定hbase在HDFS上存储的路径 -->
<property>
<name>hbase.rootdir</name>
<value>hdfs://ns1/hbase</value>
</property>
<!-- 指定hbase是分布式的 -->
<property>
<name>hbase.cluster.distributed</name>
<value>true</value>
<property>
<name>hbase.master.info.port</name>
<value>60010</value>
</property>
</property>
<!-- 指定zk的地址，多个用","分割 -->
<property>
<name>hbase.zookeeper.quorum</name>
<value>hadoop11:2181, hadoop12:2181, hadoop13:2181</value>
</property>
</configuration>
```

（5）增加 slave 的集群。

vim /usr/app/hbase-1.0.0/conf/regionservers

注意：部署到 DataNode 上面，哪一台启动 HBase 哪一台就是 Master。

```
hadoop11
hadoop12
hadoop13
```

（6）复制 HBase 到其他节点。

```
scp -r /usr/app/hbase-1.0.0 root@hadoop12:/usr/app/
scp -r /usr/app/hbase-1.0.0 root@hadoop13:/usr/app/
```

（7）将配置好的 HBase 复制到每一个节点并同步时间。

（8）启动所有的 HBase。前提需要启动 ZooKeeper 和 HDFS。

分别启动 zk：

```
./zkServer.sh start
```

启动 HBase 集群：

```
start-dfs.sh
```

启动 HBase，在主节点上运行：

```
/usr/app/hbase-1.0.0/bin/start-hbase.sh
```

（9）通过浏览器访问 HBase 管理页面。
http://192.168.200.11:60010

（10）为保证集群的可靠性，要启动多个 HMaster。

```
hbase-daemon.sh start master
```

6.3 HBase 实战

6.3.1 HBase 客户端的操作

1. 启动 HBase 集群

启动 HBase 集群，在每一台 HBase 服务器启动 HBase 服务，命令如下：

```
/usr/app/hbase-1.0.0/bin/start-hbase.sh
```

2. 启动命令

```
$HBASE_HOME/bin/hbase shell
```

如果有 kerberos 认证，需要事先使用相应的 keytab 进行一下认证（使用 kinit 命令），认证成功之后，再使用 HBase shell 进入，可以使用 whoami 命令查看当前用户。

```
hbase(main):001:0> whoami
root (auth:SIMPLE)
    groups: root
```

3. 表的管理

（1）查看有哪些表

```
hbase(main):002:0> list
TABLE
YH_BigData
1 row(s) in 0.9320 seconds

=> ["YH_BigData"]
```

（2）创建表

语法：create <table>, {NAME =><family>, VERSIONS =><VERSIONS>}

例如：创建表 t1，有两个 family name：f1，f2，且版本数均为 2。

```
hbase(main)>create 't1' , {NAME => 'f1', VERSIONS => 2}
```

（3）删除表

分两步：首先 disable，然后 drop。

例如：删除表 t1。

```
hbase(main)> disable 't1'
hbase(main)> drop 't1'
```

（4）查看表的结构

语法：describe <table>

例如：查看表 t1 的结构。

```
hbase(main):012:0> describe 't1'
Table t1 is ENABLED
t1
COLUMN FAMILIES DESCRIPTION
{NAME => 'f1', DATA_BLOCK_ENCODING => 'NONE', BLOOMFILTER => 'ROW',
REPLICATION_SCOPE => '0', VERSIONS => '2', COM
PRESSION => 'NONE', MIN_VERSIONS => '0', TTL => 'FOREVER',
KEEP_DELETED_CELLS => 'FALSE', BLOCKSIZE => '65536', IN
_MEMORY => 'false', BLOCKCACHE => 'true'}
1 row(s) in 0.0510 seconds
```

4. 表的操作

（1）添加数据

语法：put <table>,<rowkey>,<family:column>,<value>,<timestamp>

例如：给表 t1 添加一行记录：rowkey 是 rowkey001，family name：f1，column name：col1，value：value01，timestamp：系统默认。

```
hbase(main)> put 't1','rowkey001','f1:col1','value01'
hbase(main)> put 't1','rowkey001','f1:col2','value02'
hbase(main)> put 't1','rowkey001','f1:col3','value03'
```

用法比较单一。

（2）查询数据

查询某行记录。

语法：get <table>,<rowkey>,[<family:column>,....]

例如：查询表 t1，rowkey001 中的 f1 下的 col1 的值。

```
hbase(main)> get 't1','rowkey001', 'f1:col1'
# 或者：
hbase(main)> get 't1','rowkey001', {COLUMN=>'f1:col1'}
```

查询表 t1，rowke001 中的 f1 下的所有列值：

```
hbase(main)> get 't1','rowkey001'
COLUMN                    CELL
 f1:col1                  timestamp=1489008126383, value=value01
 f1:col2                  timestamp=1489008174938, value=value02
 f1:col3                  timestamp=1489008182248, value=value03
```

（3）扫描表

语法：scan <table>, {COLUMNS => [<family:column>,....], LIMIT => num}

另外，还可以添加 STARTROW、TIMERANGE 和 FITLER 等高级功能。

例如：扫描表 t1 的前 5 条数据。

```
hbase(main)> scan 't1',{LIMIT=>5}
```

（4）查询表中的数据行数

语法：count <table>, {INTERVAL => intervalNum, CACHE => cacheNum}

INTERVAL 设置多少行显示一次及对应的 rowkey，默认 1000；CACHE 每次去取的缓存区大小，默认是 10，调整该参数可提高查询速度。

例如，查询表 t1 中的行数，每 100 条显示一次，缓存区为 500。

```
hbase(main)> count 't1', {INTERVAL => 100, CACHE => 500}
```

（5）删除数据

删除行中的某个列值。

语法：delete <table>, <rowkey>, <family:column> , <timestamp>,必须指定列名

例如：删除表 t1 的 rowkey001 中的 f1:col1 的数据。

```
hbase(main)> delete 't1','rowkey001','f1:col1'
```

注：将删除 rowkey001 行中 f1:col1 列所有版本的数据。

（6）删除行

语法：deleteall <table>, <rowkey>, <family:column> , <timestamp>

可以不指定列名，删除整行数据。

例如：删除表 t1 中 rowk001 行的数据。

```
hbase(main)> deleteall 't1','rowkey001'
```

（7）删除表中的所有数据

语法： truncate <table>

其具体过程是：disable table -> drop table -> create table

例如：删除表 t1 的所有数据。

```
hbase(main)> truncate 't1'
```

（8）修改数据

语法：put <table>,<rowkey>,<family:column>,<value>,<timestamp>

例如：给表 t1 的添加一行记录：rowkey 是 rowkey001，family name：f1，column

name。

```
hbase(main):046:0> put 't1','rowkey001','f1:col2','value02'
0 row(s) in 0.0150 seconds

hbase(main):047:0> scan 't1',{LIMIT=>5}
ROW                     COLUMN+CELL
 rowkey001                column=f1:col2, timestamp=1489008436490, value=value02
1 row(s) in 0.0190 seconds

hbase(main):048:0> put 't1','rowkey001','f1:col2','value05'
0 row(s) in 0.0160 seconds

hbase(main):049:0> scan 't1',{LIMIT=>5}
ROW                     COLUMN+CELL
 rowkey001                column=f1:col2, timestamp=1489008442304, value=value05
1 row(s) in 0.0120 seconds
```

6.3.2 Java 操作 HBase

HBase 连接客户端时候，不需要将任何配置放入到 resources 中。

```java
package cn.orcale.com.bigdata.hbase;

import java.io.IOException;
import java.util.ArrayList;
import java.util.List;

import org.apache.hadoop.conf.Configuration;
import org.apache.hadoop.hbase.HBaseConfiguration;
import org.apache.hadoop.hbase.HColumnDescriptor;
import org.apache.hadoop.hbase.HTableDescriptor;
import org.apache.hadoop.hbase.KeyValue;
import org.apache.hadoop.hbase.MasterNotRunningException;
import org.apache.hadoop.hbase.ZooKeeperConnectionException;
import org.apache.hadoop.hbase.client.Delete;
import org.apache.hadoop.hbase.client.Durability;
import org.apache.hadoop.hbase.client.Get;
import org.apache.hadoop.hbase.client.HBaseAdmin;
import org.apache.hadoop.hbase.client.HTable;
import org.apache.hadoop.hbase.client.Put;
import org.apache.hadoop.hbase.client.Result;
import org.apache.hadoop.hbase.client.ResultScanner;
import org.apache.hadoop.hbase.client.Scan;
import org.apache.hadoop.hbase.util.Bytes;
import org.apache.log4j.Logger;
/**
 *
 * ClassName: HbaseUtil
```

```java
 * @Description: HBase 交互工具类
 * @author 余辉
 * @date 2017-3-18
 */
public class HbaseDao {

    public static Logger log = Logger.getLogger(HbaseDao.class);

    public static Configuration configuration;
    static {
        configuration = HBaseConfiguration.create();
        configuration.set("hbase.zookeeper.quorum", "hadoop11, hadoop12, hadoop13");
        configuration.set("hbase.zookeeper.property.clientPort", "2181");
    }

    public static void main(String[] args) {
        createTable("YH_BigData");
//        dropTable("YH_BigData");
//
//        insertData("xiaoming" , "YH_BigData");
//
//        QueryAll("YH_BigData");
//        QueryByCondition1("xiaohui",  "YH_BigData");
//        QueryByCondition1("xiaoming",  "YH_BigData");
//
//        deleteRow("YH_BigData","xiaoming");

    }

    /**
     * 创建表
     * @param tableName
     */
    public static void createTable(String tableName) {
        log.info("start create table ......");
        try {
            HBaseAdmin hBaseAdmin = new HBaseAdmin(configuration);
            if (hBaseAdmin.tableExists(tableName)) {// 如果存在要创建的表，那么先删除，再创建
                hBaseAdmin.disableTable(tableName);
                hBaseAdmin.deleteTable(tableName);
                log.info("-------------->"+tableName + " is exist,detele....");
            }
            HTableDescriptor tableDescriptor = new HTableDescriptor(tableName);
            tableDescriptor.addFamily(new HColumnDescriptor("column1"));
```

```java
            tableDescriptor.addFamily(new
HColumnDescriptor("column2"));
            tableDescriptor.addFamily(new
HColumnDescriptor("column3"));
            hBaseAdmin.createTable(tableDescriptor);

        } catch (MasterNotRunningException e) {
            e.printStackTrace();
        } catch (ZooKeeperConnectionException e) {
            e.printStackTrace();
        } catch (IOException e) {
            e.printStackTrace();
        }
        log.info("end create table ......");
    }

    /**
     * 插入数据
     * @param tableName
     */
    private static void insertData(String rowkey , String tableName) {
        try {
            HTable ht = new HTable(configuration, tableName);

            //RowKey
            Put put = new Put(rowkey.getBytes());

            //            列族            列名            值
            put.add(Bytes.toBytes("column2"),
"screen_width".getBytes(), "1080".getBytes());// 本行数据的第一列
            put.add(Bytes.toBytes("column2"),
"screen_height".getBytes(), "1920".getBytes());// 本行数据的第二列
            put.add(Bytes.toBytes("column2"), "url".getBytes(),
"www.baidu.com".getBytes());// 本行数据的第三列
            put.add(Bytes.toBytes("column2"), "event_data".getBytes(),
"12|16|13|17|12|16".getBytes());// 本行数据的第四列

            put.setDurability(Durability.SYNC_WAL);

            //put a record
            ht.put(put);
            ht.close();
        } catch (IOException e) {
            e.printStackTrace();
        }



    }
```

```java
    /**
     * 删除一张表
     * @param tableName
     */
    public static void dropTable(String tableName) {
        try {
            HBaseAdmin admin = new HBaseAdmin(configuration);
            admin.disableTable(tableName);
            admin.deleteTable(tableName);
            log.info("-------------->"+tableName + " is exist,detele....");
        } catch (MasterNotRunningException e) {
            e.printStackTrace();
        } catch (ZooKeeperConnectionException e) {
            e.printStackTrace();
        } catch (IOException e) {
            e.printStackTrace();
        }

    }
    /**
     * 根据 rowkey 删除一条记录
     * @param tablename
     * @param rowkey
     */
    public static void deleteRow(String tablename, String rowkey)  {
        try {
            HTable table = new HTable(configuration, tablename);
            List list = new ArrayList();
            Delete d1 = new Delete(rowkey.getBytes());
            list.add(d1);

            table.delete(list);
            System.out.println("-------------->"+"删除行成功!");

        } catch (IOException e) {
            e.printStackTrace();
        }


    }

    /**
     * 查询所有数据
     * @param tableName
     */
    public static void QueryAll(String tableName) {
        try {
```

```java
        HTable table = new HTable(configuration, tableName);
        ResultScanner rs = table.getScanner(new Scan());
        for (Result r : rs) {
            System.out.println("获得到rowkey:" + new String(r.getRow()));
            for (KeyValue keyValue : r.raw()) {
                System.out.println("列: " + new String(keyValue.getFamily())
                        + "====值:" + new String(keyValue.getValue()));
            }
        }
    } catch (IOException e) {
        e.printStackTrace();
    }
}

/**
 * 单条件查询,根据rowkey查询唯一一条记录
 * @param tableName
 */
public static void QueryByCondition1(String rowkey , String tableName) {

    try {
    HTable table = new HTable(configuration, tableName);
        Get scan = new Get(rowkey.getBytes());// 根据rowkey查询
        Result r = table.get(scan);
        System.out.println("获得到rowkey:" + new String(r.getRow()));
        for (KeyValue keyValue : r.raw()) {
            System.out.println("列: " + new String(keyValue.getFamily())
                    + "====值:" + new String(keyValue.getValue()));
        }
    } catch (IOException e) {
        e.printStackTrace();
    }
}
}
```

6.3.3 Scala 操作 HBase

```
package com.ou.cn.hbase

import org.apache.hadoop.hbase.client._
import org.apache.hadoop.hbase.util.Bytes
import org.apache.hadoop.hbase.{HBaseConfiguration, HColumnDescriptor,
```

```scala
  HTableDescriptor, TableName}
/**
  * Created by yuhui on 2017/6/1 0001.
  */
object hbaseCli {
  //创建表
  def createHTable(connection: Connection,tablename: String): Unit=
  {
    //HBase 表模式管理器
    val admin = connection.getAdmin
    //本例将操作的表名
    val tableName = TableName.valueOf(tablename)
    //如果需要创建表
    if (!admin.tableExists(tableName)) {
      //创建 HBase 表模式
      val tableDescriptor = new HTableDescriptor(tableName)
      //创建列簇1    artitle
      tableDescriptor.addFamily(new HColumnDescriptor("artitle".getBytes()))
      //创建列簇2    author
      tableDescriptor.addFamily(new HColumnDescriptor("author".getBytes()))
      //创建表
      admin.createTable(tableDescriptor)
      println("create done.")
    }
  }

  //删除表
  def deleteHTable(connection:Connection,tablename:String):Unit={
    //本例将操作的表名
    val tableName = TableName.valueOf(tablename)
    //HBase 表模式管理器
    val admin = connection.getAdmin
    if (admin.tableExists(tableName)){
      admin.disableTable(tableName)
      admin.deleteTable(tableName)
    }
  }

  //插入记录
  def insertHTable(connection:Connection,tablename:String,family:String,column:String,key:String,value:String):Unit={
    try{
      val userTable = TableName.valueOf(tablename)
      val table=connection.getTable(userTable)
      //准备 key 的数据
      val p=new Put(key.getBytes)
```

```scala
      //为put操作指定column和value
      p.addColumn(family.getBytes,column.getBytes,value.getBytes())
      //提交一行
      table.put(p)
    }
  }
  //基于KEY查询某条数据
  def getAResult(connection:Connection,tablename:String,family:String,column:String,key:String):Unit={
    var table:Table=null
    try{
      val userTable = TableName.valueOf(tablename)
      table=connection.getTable(userTable)
      val g=new Get(key.getBytes())
      val result=table.get(g)
      val value=Bytes.toString(result.getValue(family.getBytes(),column.getBytes()))
      println("key:"+value)
    }finally{
      if(table!=null)table.close()

    }

  }
  //删除某条记录
  def deleteRecord(connection:Connection,tablename:String,family:String,column:String,key:String): Unit ={
    var table:Table=null
    try{
      val userTable=TableName.valueOf(tablename)
      table=connection.getTable(userTable)
      val d=new Delete(key.getBytes())
      d.addColumn(family.getBytes(),column.getBytes())
      table.delete(d)
      println("delete record done.")
    }finally{
      if(table!=null)table.close()
    }
  }

  //扫描记录
  def scanRecord(connection:Connection,tablename:String,family:String,column:String): Unit ={
    var table:Table=null
    var scanner:ResultScanner=null
    try{
      val userTable=TableName.valueOf(tablename)
```

```scala
      table=connection.getTable(userTable)
      val s=new Scan()
      s.addColumn(family.getBytes(),column.getBytes())
      scanner=table.getScanner(s)
      println("scan...for...")
      var result:Result=scanner.next()
      while(result!=null){
        println("Found row:" + result)
        println("Found value: "+Bytes.toString(result.getValue(family.getBytes(),column.getBytes())))
        result=scanner.next()
      }


    }finally{
      if(table!=null)
        table.close()
      scanner.close()
    }
  }

  def main(args: Array[String]): Unit = {
    // val sparkConf = new SparkConf().setAppName("HBaseTest")
    //启用spark上下文，只有这样才能驱动spark并行计算框架
    //val sc = new SparkContext(sparkConf)
    //创建一个配置，采用的是工厂方法
    val conf = HBaseConfiguration.create
    conf.set("hbase.zookeeper.property.clientPort", "2181")
    conf.set("hbase.zookeeper.quorum", "hadoop11, hadoop12, hadoop13")

    try{
      //Connection 的创建是个重量级的工作，线程安全，是操作 HBase 的入口
      val connection= ConnectionFactory.createConnection(conf)
      //创建表测试
      try {
//        createHTable(connection, "HadoopAndSpark")
        //插入数据,重复执行为覆盖
//        insertHTable(connection,"HadoopAndSpark","artitle","Hadoop","002","Hadoop for me")
//        insertHTable(connection,"HadoopAndSpark","artitle","Hadoop","003","Java for me")
//        insertHTable(connection,"HadoopAndSpark","artitle","Spark","002","Scala for me")
```

```
            //删除记录
//        deleteRecord(connection,"HadoopAndSpark","artitle","Spark","002")

            //扫描整个表
//          scanRecord(connection,"HadoopAndSpark","artitle","Hadoop")

            getAResult(connection,"HadoopAndSpark","artitle","Hadoop",
"002")

            //删除表测试
//          deleteHTable(connection, "HadoopAndSpark")
        }finally {
            connection.close
            //   sc.stop
        }
    }
  }
}
```

6.4 小结

  HBase 是一个分布式的、面向列的开源数据库。本章讲解 HBase 组件：首先介绍 HBase 简介和架构，带领读者熟悉 HBase；接着描述如何安装 HBase 集群；最后实战方面通过 HBase 客户端、Eclipse+Java、Intellij idea+Scala 对 HBase 进行操作。

第 7 章 ◀ Hive实战 ▶

7.1 Hive 介绍和架构

Hive 是基于 Hadoop 的一个数据仓库工具,可以将结构化的数据文件映射为一张数据库表,并提供完整的 SQL 查询功能,可以将 SQL 语句转换为 MapReduce 任务进行运行。其优点是学习成本低,可以通过类 SQL 语句快速实现简单的 MapReduce 统计,不必开发专门的 MapReduce 应用,十分适合数据仓库的统计分析。

Hive 是建立在 Hadoop 上的数据仓库基础构架。它提供了一系列的工具,可以用来进行数据提取转化加载(ETL),这是一种可以存储、查询和分析存储在 Hadoop 中的大规模数据的机制。Hive 定义了简单的类 SQL 查询语言,称为 HQL,它允许熟悉 SQL 的用户查询数据。同时,这个语言也允许熟悉 MapReduce 开发者开发自定义的 mapper 和 reducer 来处理内建的 mapper 和 reducer 无法完成的、复杂的分析工作。

1. Hive 与关系数据库的区别

(1)Hive 和关系数据库存储文件的系统不同,Hive 使用的是 Hadoop 的 HDFS(Hadoop 的分布式文件系统),关系数据库则是服务器本地的文件系统。

(2)Hive 使用的计算模型是 MapReduce,而关系数据库则是自己设计的计算模型。

(3)关系数据库都是为实时查询的业务进行设计的,而 Hive 则是为海量数据做数据挖掘设计的,实时性很差;实时性的区别导致 Hive 的应用场景和关系数据库有很大的不同。

(4)Hive 很容易扩展自己的存储能力和计算能力,这个是继承 Hadoop 的,而关系数据库在这个方面要比 Hive 差很多。

2. Hive 架构介绍

Hive 架构如图 7-1 所示。

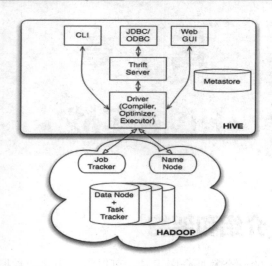

图 7-1 Hive 架构图

Hive 的体系结构可以分为以下几部分：

（1）用户接口主要有三个：CLI、Client 和 WUI。其中最常用的是 CLI，CLI 启动的时候，会同时启动一个 Hive 副本。Client 是 Hive 的客户端，用户连接至 Hive Server。在启动 Client 模式的时候，需要指出 Hive Server 所在节点，并且在该节点启动 Hive Server。WUI 是通过浏览器访问 Hive 的。

（2）Hive 将元数据存储在数据库中，如 MySQL、Derby。Hive 中的元数据包括表的名字、表的列、表的分区、表的属性（是否为外部表等）、表的数据所在目录等。

（3）解释器、编译器、优化器完成 HQL 查询语句。生成的查询计划存储在 HDFS 中，并在随后有 MapReduce 调用执行。

（4）Hive 的数据存储在 HDFS 中，大部分的查询、计算由 MapReduce 完成（包含*的查询，比如 select*from tbl 不会生成 MapRedcue 任务）。Hive 将元数据存储在 RDBMS 中。

3. 数据准备

数据文件内容如表 7-1 所示。

表 7-1 数据文件内容

people01.txt	people02.txt
1,man,baba	1,women,mama
2,man,jiujiu	2,women,laopo
3,man,yuhui	3,women,jiejie
4,man,huangyujia	4,women,xiaoyuyan
5,man,xiaoting	5,women,jiuma

7.2 Hive 数据类型和表结构

Hive 的内置数据类型可以分为两大类：

- 基础数据类型。
- 复杂数据类型。

1. 基础数据类型

基础数据类型包括：TINYINT、SMALLINT、INT、BIGINT、BOOLEAN、FLOAT、DOUBLE、STRING、BINARY、TIMESTAMP、DECIMAL、CHAR、VARCHAR、DATE。Hive 基础数据类型如图 7-2 所示，下面的表格列出这些基础类型所占的字节以及 Hive 从什么版本开始支持这些类型。

数据类型	所占字节	开始支持版本
TINYINT	1byte，-128～127	
SMALLINT	2byte，-32,768～32,767	
INT	4byte,-2,147,483,648～2,147,483,647	
BIGINT	8byte,-9,223,372,036,854,775,808～9,223,372,036,854,775,807	
BOOLEAN		
FLOAT	4byte单精度	
DOUBLE	8byte双精度	
STRING		
BINARY		从Hive0.8.0开始支持
TIMESTAMP		从Hive0.8.0开始支持
DECIMAL		从Hive0.11.0开始支持
CHAR		从Hive0.13.0开始支持
VARCHAR		从Hive0.12.0开始支持
DATE		从Hive0.12.0开始支持

图 7-2 Hive 基础数据类型

2. 复杂数据类型

复杂类型包括 ARRAY、MAP、STRUCT、UNION，这些复杂类型是由基础类型组成的。

（1）ARRAY：ARRAY 类型是由一系列相同数据类型的元素组成，这些元素可以通过下标来访问。比如有一个 ARRAY 类型的变量 fruits，它是由 ['apple','orange','mango']组成，那么我们可以通过 fruits[1]来访问元素 orange，因为 ARRAY 类型的下标是从 0 开始的。

（2）MAP：MAP 包含 key->value 键值对，可以通过 key 来访问元素。例如：访问元素，假设 user 是一个 map 类型，其中 username 是 key，password 是 value；那么我们可以通过 userlist['username']来得到这个用户对应的 password。

（3）STRUCT：STRUCT 可以包含不同数据类型的元素，可以通过"点语法"的方式来得到所需要的元素，比如 user 是一个 STRUCT 类型，那么可以通过 user.address 得到这个用户的地址。

（4）UNION：UNIONTYPE，这个数据类型是从 Hive 0.7.0 开始支持的。

3. 内部表和外部表的区别

内部表和外部表的区别如图 7-3 所示。

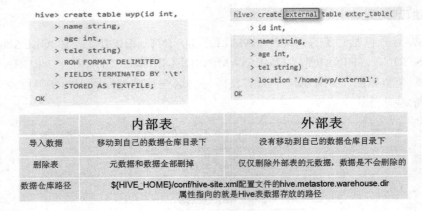

图 7-3 内部表和外部表的区别

7.3 Hive 分区、桶与倾斜

1. Hive 的分区

（1）在 Hive Select 查询中一般会扫描整个表内容，会消耗很多时间做没必要的工作。有时候查询只需要扫描表中关心的一部分数据，因此建表时引入了 partition 概念。

（2）分区表指的是在创建表时指定的 partition 的分区空间。

（3）如果需要创建有分区的表，需要在 create 表的时候调用可选参数 partitioned by，详见表创建的语法结构。

（4）一个表可以拥有一个或者多个分区，每个分区以文件夹的形式单独存储在表文件夹的目录下。

（5）表和列名不区分大小写。

（6）分区是以字段的形式在表结构中存在，通过 describe table 命令可以查看到字段存在，但是该字段不存放实际的数据内容，仅仅是分区的表示。

2. Hive 桶表

分桶其实就是把大表化成了"小表"，然后通过两个表相同列使用 Map 端连接（Map-Side Join），这是用来解决大表与小表之间的连接问题。将桶中的数据按某列进行排序会提高查询效率。

（1）创建分桶表

```
CREATE TABLE tb_bucket_shop (
    shop_id int,
    shop_name string,
    shopkeeper string
) CLUSTERED BY (shop_id) INTO 4BUCKETS
ROW FORMAT DELIMITED
FIELDS TERMINATED BY '\t';
```

（2）数据导入

```
set hive.enforce.bucketing=true;
INSERT OVERWRITE TABLE tb_bucket_shop SELECT shop_id, shop_name, shopkeeper FROM tb_part_shop CLUSTER BY shop_id;
```

若没有使用 hive.enforce.bucketing 属性，则需要设置和分桶个数相匹配的 reducer 个数，同时 SELECT 后添加 CLUSTER BY

```
set mapred.reduce.tasks=4;
INSERT OVERWRITE TABLE tb_bucket_shop SELECT shop_id, shop_name, shopkeeper FROM tb_part_shop CLUSTER BY shop_id;
```

分桶适合于 sampling，不过其数据正确地导入到 hive 表中，需要用户自己来保证，因为 table 中信息仅仅是元数据，而不影响实际填充表的命令。

（3）小结

分区分桶是 hive 性能优化的一个手段，不同的字段，其数值属性不同，其对应的优化方式也不同。不能简单地认为分区分桶对应传统关系型数据库的分库分表，两者完全不一样。

3. Hive 数据倾斜

（1）数据倾斜的原因

使 map 的输出数据更均匀地分布到 Reduce 中，是我们的最终目标。由于 Hash 算法的局限性，按 key Hash 会或多或少地造成数据倾斜。大量经验表明数据倾斜的原因是人为的建表疏忽或业务逻辑，是可以规避的。

（2）解决思路

Hive 的执行是分阶段的，map 处理数据量的差异取决于上一个 stage 的 reduce 输出，所以如何将数据均匀地分配到各个 reduce 中，就是解决数据倾斜的根本所在。

（3）具体办法

- 内存优化和 I/O 优化。
- 驱动表：使用大表做驱动表，以防止内存溢出；Join 最右边的表是驱动表；Mapjoin 无视 join 顺序，用大表做驱动表。

4. Mapjoin 是一种避免数据倾斜的手段

允许在 map 阶段进行 join 操作，MapJoin 把小表全部读入内存中，在 map 阶段直接拿另外一个表的数据和内存中表数据做匹配。由于对 map 进行了 join 操作，省去了 reduce 运行，效率也会高很多。

在对多个表 join 连接操作时，将小表放在 join 的左边，大表放在 jion 的右边。在执行这样的 join 连接时，小表中的数据会被缓存到内存当中，这样可以有效减少发生内存溢出错误的几率。

（1）设置参数

```
hive.map.aggr = true
hive.groupby.skewindata=true
```

还有其他参数，请读者自行查阅官网相关文档。

（2）SQL 语言调节

比如：在 group by 维度过小时，采用 sum() group by 的方式来替换 count(distinct)完成计算。

（3）StreamTable

将在 reducer 中进行 join 操作时的小 table 放入内存，而大 table 通过 stream 方式读取。

（4）索引

Hive 从 0.80 开始才有索引，它提供了一个 Bitmap 位图索引，索引可以加快 group by 查询语句的执行速度，但用得较少。

7.4 Hive 安装

1. Hive 安装步骤

（1）上传 hive-0.12.0 压缩包，解压。
（2）安装 MySQL 服务器。
（3）进入 Hive 的 conf 目录新建一个 hive-site.xml。
（4）在 hive-site.xml 中写入 MySQL 连接信息。
（5）将 MySQL 的驱动包复制到 Hive 的 lib 目录下 app/hive-0.12.0/lib。
（6）执行 sh /usr/app/hive-0.12.0/bin/hive，启动 Hive。

2. 配置环境变量

vi /etc/profile

```
export JAVA_HOME=/usr/app/jdk1.7
export HADOOP_HOME=/usr/app/hadoop-2.6.0
export HBASE_HOME=/usr/app/hbase-1.0.0
```

```
export HIVE_HOME=/usr/app/hive-0.12.0
export
PATH=$PATH:$JAVA_HOME/bin:$HADOOP_HOME/bin:$HBASE_HOME:$HIVE_HOME/bin
```

3. Hive 启动配置添加内容

修改/usr/app/hive-0.12.0/conf/hive-env.sh 的尾部 hive-env.sh，增加以下三行。

```
export JAVA_HOME=/usr/app/jdk1.7
export HADOOP_HOME=/usr/app/hadoop-2.6.0
export HIVE_HOME=/usr/app/hive-0.12.0
```

4. 修改主要配置文件 hive-site.xml

将 hive-default.xml.template 另存为 hive-site.xml，再修改配置文件 /usr/app/hive-0.12.0/conf/hive-site.xml。

```
<configuration>
<property>
<name>javax.jdo.option.ConnectionURL</name>

    <value>jdbc:mysql://hadoop11:3306/hive?createDatabaseIfNotExist=true</value>
      <description>JDBC connect string for a JDBC
metastore</description>
</property>

<property>
      <name>javax.jdo.option.ConnectionDriverName</name>
         <value>com.mysql.jdbc.Driver</value>
      <description>Driver class name for a JDBC
metastore</description>
</property>

<property>
      <name>javax.jdo.option.ConnectionUserName</name>
         <value>root</value>
      <description>username to use against metastore
database</description>
</property>

<property>
      <name>javax.jdo.option.ConnectionPassword</name>
         <value>123456</value>
      <description>password to use against metastore
database</description>
</property>
<property>
```

```
<name>hive.server2.thrift.sasl.qop</name>
<value>auth</value>
</property>
<property>
<name>hive.metastore.schema.verification</name>
<value>false</value>
</property>
</configuration>
```

5. 启动 Hive

```
sh /usr/app/hive-0.12.0/bin/hive

hive> create table test(id int,name string);
OK
Time taken: 8.292 seconds
hive> show tables;
OK
```

6. 查看 Hive 在 HDFS 创建目录

```
[root@hadoop13 ~]# hadoop fs -lsr /
drwxr-xr-x   - root supergroup          0 2016-01-10 20:57 /user
drwxr-xr-x   - root supergroup          0 2016-01-10 20:57 /user/hive
drwxr-xr-x   - root supergroup          0 2016-01-11 01:46 /user/hive/warehouse
   drwxr-xr-x   - root supergroup          0 2016-01-11 01:46 /user/hive/warehouse/test
```

7.5 Hive 实战

7.5.1 Hive 客户端的操作

1. Hive 的启动

```
[root@hadoop11 bin]# ls
beeline  ext  hive  hive-config.sh  hiveserver2  metatool  schematool
[root@hadoop11 bin]# pwd
/usr/app/hive-0.12.0/bin
[root@hadoop11 bin]# Sh /usr/app/hive-0.12.0/bin/hive
```

2. Hive 创建数据库

Hive 是一种数据库技术，可以定义数据库和表来分析结构化数据。主题结构化数据分析是以表方式存储数据，并通过查询来分析。本章介绍如何创建 Hive 数据库，配置单元包含一个名为 default 默认的数据库。

在 Hive 中，数据库是一个命名空间或表的集合。CREATE DATABASE 语句是 Hive 中用来创建数据库的语句。此语句语法声明如下：

```
CREATE DATABASE|SCHEMA [IF NOT EXISTS]<database name>
```

在这里，IF NOT EXISTS 是一个可选子句，通知用户已经存在相同名称的数据库。在 DATABASE 的这个命令中也可以使用 SCHEMA。下面的查询执行创建一个名为 userdb 数据库：

```
hive> CREATE DATABASE [IF NOT EXISTS] userdb;
hive> CREATE DATABASE userdb;
```

或

```
hive> CREATE SCHEMA userdb;
```

下面的查询用于验证数据库列表：

```
hive>SHOW DATABASES;
default
userdb
```

进入数据库：

```
hive>use default;
OK
Time taken: 0.114 seconds
hive>show tables;
OK
people
test
Time taken: 0.404 seconds, Fetched: 2 row(s)
hive>
```

3. Hive 删除数据库

DROP DATABASE 是删除所有的表并删除数据库的语句。它的语法如下：

```
DROP DATABASE StatementDROP (DATABASE|SCHEMA) [IF EXISTS]
database_name
[RESTRICT|CASCADE];
```

下面的查询用于删除数据库。假设要删除的数据库名称为 userdb。

```
hive>DROP DATABASE IF EXISTS userdb;
```

以下是使用 CASCADE 查询删除数据库。这意味着要全部删除相应的表在删除数据库之前。

```
hive> DROP DATABASE IF EXISTS userdb CASCADE;
```

以下使用 SCHEMA 查询删除数据库。

```
hive> DROP SCHEMA userdb;
```

4. 添加数据及创建 Hive 表

（1）数据展示

数据文件内容如表 7-2 所示。Linux 中查看数据文件内容如图 7-4 所示。

表 7-2 数据文件内容

people01.txt	people02.txt
1,man,baba 2,man,jiujiu 3,man,yuhui 4,man,huangyujia 5,man,xiaoting	1,women,mama 2,women,laopo 3,women,jiejie 4,women,xiaoyuyan 5,women,jiuma

```
[root@hadoop11 ~]# cat people01.txt
1,man,baba
2,man,jiujiu
3,man,yuhui
4,man,huangyujia
5,man,xiaoting
[root@hadoop11 ~]# cat people02.txt
1,women,mama
2,women,laopo
3,women,jiejie
4,women,xiaoyuyan
5,women,jiuma
[root@hadoop11 ~]#
```

图 7-4 Linux 中查看文件内容

（2）建立文件

```
hadoop fs -mkdir /usr/yuhui/hive
hadoop fs -mkdir /usr/yuhui/hive/20170309
hadoop fs -mkdir /usr/yuhui/hive/20170309/00
hadoop fs -mkdir /usr/yuhui/hive/20170310
hadoop fs -mkdir /usr/yuhui/hive/20170310/00
```

（3）上传文件

```
hadoop fs -put people01.txt /usr/yuhui/hive/20170309/00/people01.txt
hadoop fs -put people02.txt /usr/yuhui/hive/20170310/00/people02.txt
```

（4）HDFS 路径及文件展示（如图 7-5 所示）

/usr/yuhui/hive　Go!

Permission	Owner	Group	Size	Replication	Block Size	Name
drwxr-xr-x	root	supergroup	0 B	0	0 B	20170309
drwxr-xr-x	root	supergroup	0 B	0	0 B	20170310

/usr/yuhui/hive/20170309/00　Go!

Permission	Owner	Group	Size	Replication	Block Size	Name
-rw-r--r--	root	supergroup	72 B	3	128 MB	people01.txt

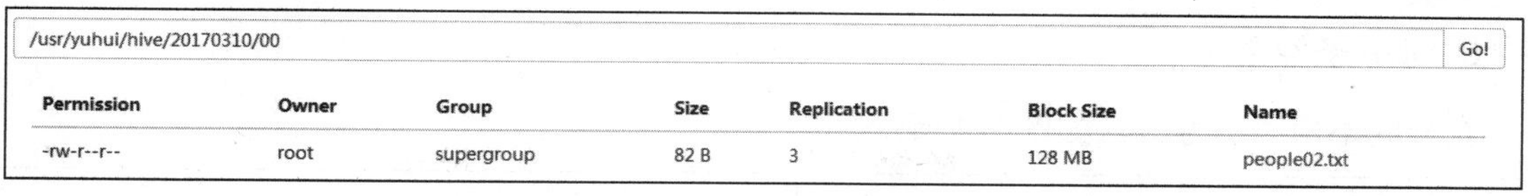

Permission	Owner	Group	Size	Replication	Block Size	Name
-rw-r--r--	root	supergroup	82 B	3	128 MB	people02.txt

路径:/usr/yuhui/hive/20170310/00

图 7-5 HDFS 路径及文件展示

（5）建立 Hive 和 HDFS 的外部关联表

```
CREATE EXTERNAL TABLE people(id int,sex string,name string)
partitioned by (logdate string,hour string) row format delimited
fields terminated by ',';
hive> show tables;
hive> select * from people;
```

（6）追加数据

```
ALTER TABLE people ADD IF NOT EXISTS
PARTITION(logdate=20170309,hour=00)LOCATION
'/usr/yuhui/hive/20170309/00';

ALTER TABLE people ADD IF NOT EXISTS
PARTITION(logdate=20170310,hour=00)LOCATION
'/usr/yuhui/hive/20170310/00';
```

5. Hive 的分区数据查看

Hive 的分区数据查看如图 7-6 所示。

```
hive> select * from people;
OK
1    man baba       20170309    00
2    man jiujiu     20170309    00
3    man yuhui      20170309    00
4    man huangyujia 20170309    00
5    man xiaoting   20170309    00
1    women mama     20170310    00
2    women laopo    20170310    00
3    women jiejie   20170310    00
4    women xiaoyuyan 20170310   00
5    women jiuma    20170310    00
Time taken: 0.635 seconds, Fetched: 10 row(s)
```

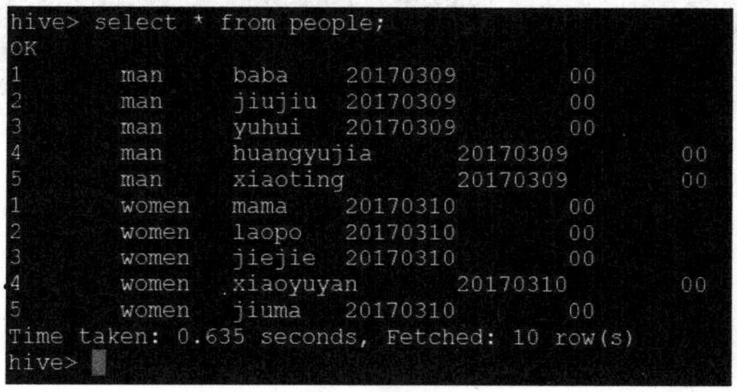

图 7-6 Hive 的分区数据查看

6. Hive 条件查询

查分区不用"引号",如图 7-7 所示。

```
hive> select * from people where logdate=20170310;
OK
1       women   mama    20170310        00
2       women   laopo   20170310        00
3       women   jiejie  20170310        00
4       women   xiaoyuyan       20170310        00
5       women   jiuma   20170310        00
Time taken: 0.324 seconds, Fetched: 5 row(s)
```

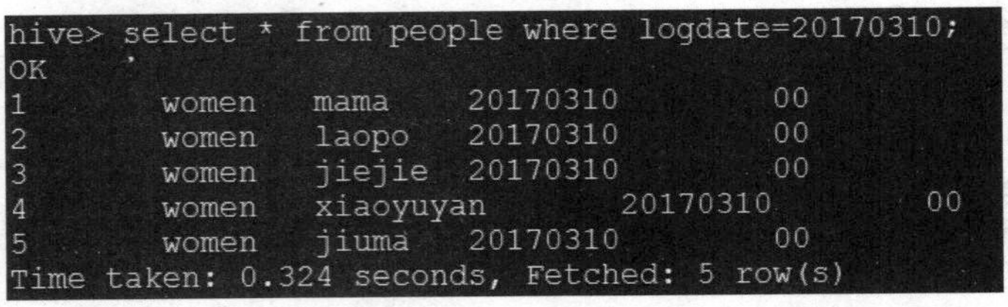

图 7-7 查分区不用"引号"示例

查条件用"引号",如图 7-8 所示。

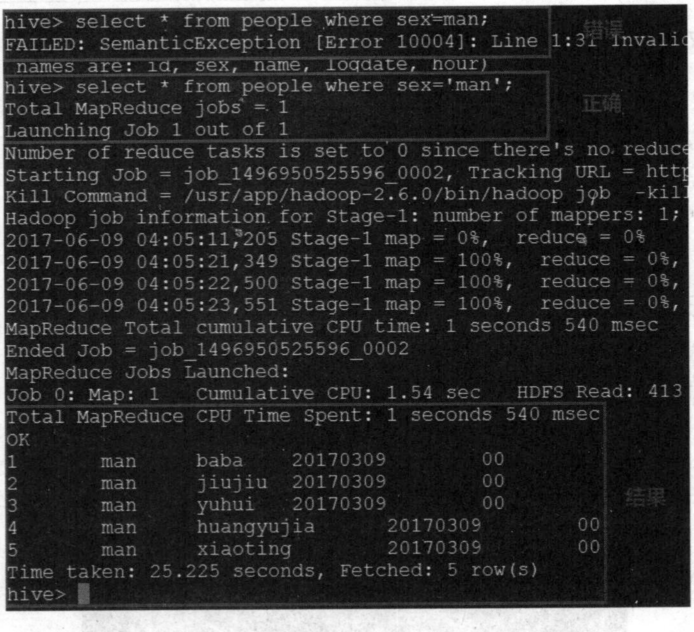

图 7-8 查条件用"引号"示例

7. UDF 函数

```
package cn.orcale.com.bigdata.hive;

import org.apache.hadoop.hive.ql.exec.UDF;
```

```java
public class helloUDF extends UDF {

    public String evaluate(String sex , String name) {

        try {

            if(sex.equals("man")){
                return sex +"  ===>  "+ name;
            }

            return sex +"  ===>  "+ name;

        }
        catch (Exception e) {
            return null;
        }
    }
}
```

Hive 中执行步骤如下，UDF 使用步骤如图 7-9 所示。

```
ADD JAR helloUDF.jar;
create temporary function helloworld as
'cn.orcale.com.bigdata.hive.helloUDF';
select helloworld(people.sex, people.name) from people ;
```

```
hive> ADD JAR helloUDF.jar;
Added helloUDF.jar to class path
Added resource: helloUDF.jar
hive> create temporary function helloworld as 'cn.orcale.com.bigdata.hive.helloUDF';
OK
Time taken: 0.553 seconds
hive> select helloworld(people.sex, people.name) from people ;
Total MapReduce jobs = 1
Launching Job 1 out of 1
Number of reduce tasks is set to 0 since there's no reduce operator
Starting Job = job_1496950525596_0003, Tracking URL = http://hadoop11:8088/proxy/application_1496950525596_0003/
Kill Command = /usr/app/hadoop-2.6.0/bin/hadoop job  -kill job_1496950525596_0003
Hadoop job information for Stage-1: number of mappers: 1; number of reducers: 0
2017-06-09 04:10:42,092 Stage-1 map = 0%,  reduce = 0%
2017-06-09 04:10:55,821 Stage-1 map = 100%,  reduce = 0%, Cumulative CPU 1.59 sec
2017-06-09 04:10:56,872 Stage-1 map = 100%,  reduce = 0%, Cumulative CPU 1.59 sec
MapReduce Total cumulative CPU time: 1 seconds 590 msec
Ended Job = job_1496950525596_0003
MapReduce Jobs Launched:
Job 0: Map: 1   Cumulative CPU: 1.59 sec   HDFS Read: 413 HDFS Write: 192 SUCCESS
Total MapReduce CPU Time Spent: 1 seconds 590 msec
OK
man    ===>  baba
man    ===>  jiujiu
man    ===>  yuhui
man    ===>  huangyujia
man    ===>  xiaoting
women  ===>  mama
women  ===>  laopo
women  ===>  jiejie
women  ===>  xiaoyuyan
women  ===>  jiuma
Time taken: 42.827 seconds, Fetched: 10 row(s)
hive>
```

图 7-9　UDF 使用步骤

8. 删除 Hive 表

```
hive> drop table people;
OK
Time taken: 1.593 seconds
hive> SHOW TABLES;
OK
test
```

9. 查询结果保存到本地

查询结果保存到本地如图 7-10、图 7-11 所示。

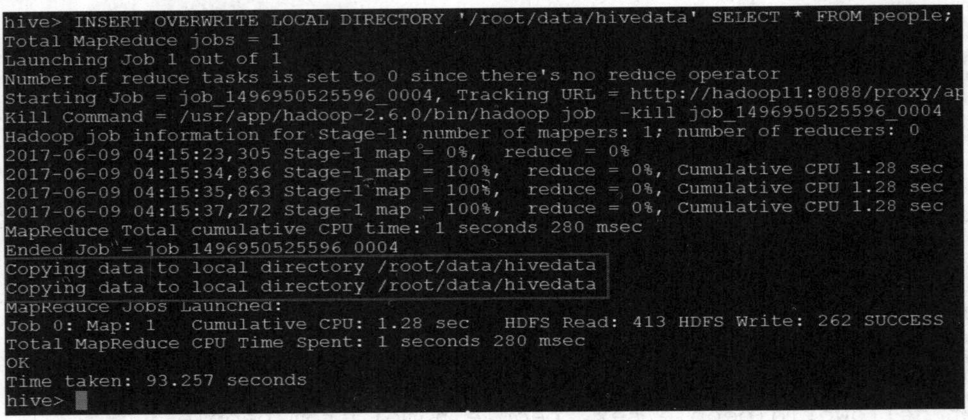

```
hive> INSERT OVERWRITE LOCAL DIRECTORY '/root/data/hivedata' SELECT * FROM people;
```

图 7-10 结果保存到本地

图 7-11 结果保存到本地

7.5.2 Hive 常用命令

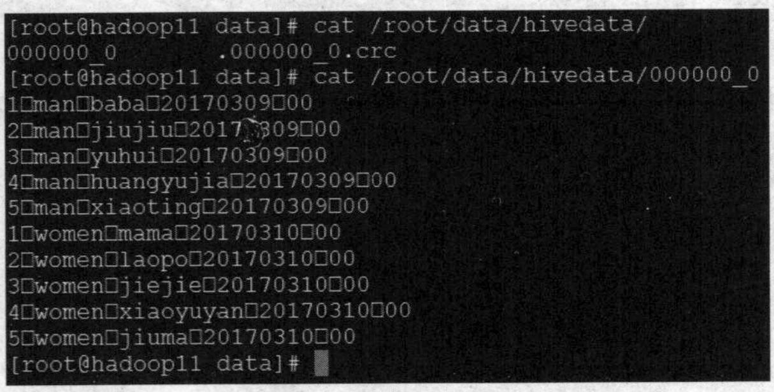

1. 创建新表

```
hive> CREATE TABLE t_hive (a int, b int, c int) ROW FORMAT DELIMITED
FIELDS TERMINATED BY '\t';
```

2. 导入数据 t_hive.txt 到 t_hive 表

```
hive> LOAD DATA LOCAL INPATH '/home/cos/demo/t_hive.txt' OVERWRITE INTO TABLE t_hive ;
```

3. 正则匹配表名

```
hive>show tables 't*';
```

4. 增加一个字段

```
hive> ALTER TABLE t_hive ADD COLUMNS (new_col String);
```

5. 重命令表名

```
hive> ALTER TABLE t_hive RENAME TO t_hadoop;
```

6. 从 HDFS 加载数据

```
hive> LOAD DATA INPATH '/user/hive/warehouse/t_hive/t_hive.txt' OVERWRITE INTO TABLE t_hive2;
```

7. 从其他表导入数据

```
hive> INSERT OVERWRITE TABLE t_hive2 SELECT * FROM t_hive ;
```

8. 创建表并从其他表导入数据

```
hive> CREATE TABLE t_hive AS SELECT * FROM t_hive2 ;
```

9. 仅复制表结构不导数据

```
hive> CREATE TABLE t_hive3 LIKE t_hive;
```

10. 通过 Hive 导出到本地文件系统

```
hive> INSERT OVERWRITE LOCAL DIRECTORY '/tmp/t_hive' SELECT * FROM t_hive;
```

11. Hive 查询 HiveQL

```
from ( select b,c as c2 from t_hive) t select t.b, t.c2 limit 2;
select b,c from t_hive limit 2;
```

7.5.3 Java 操作 Hive

1. 添加 Hive 组件下的 lib 包

连接 Hive 版本的就要在 Hive 的 lib 下面找出下列包，添加到项目中。Hive 需要的 lib 包如图 7-12 所示。

图 7-12　Hive 需要的 lib 包

2. 添加 jar 包步骤

添加 jar 包步骤如图 7-13 和图 7-14 所示。

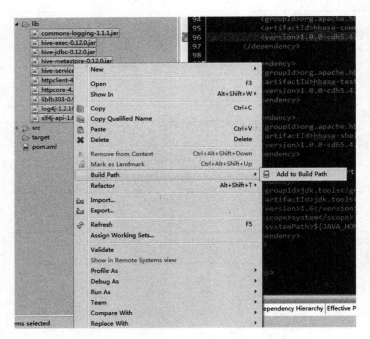

图 7-13　添加 jar 包步骤一 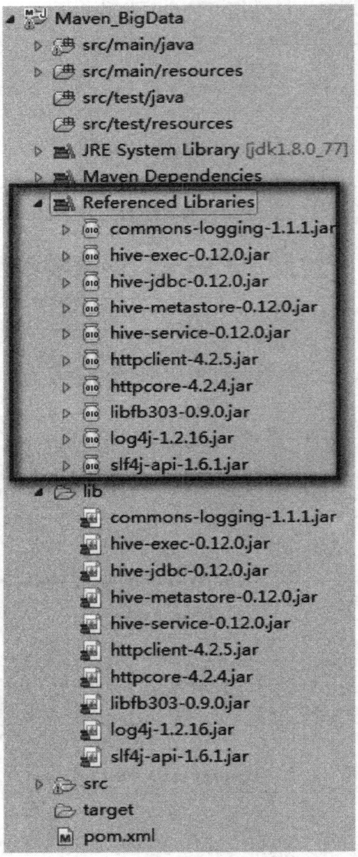　　　　图 7-14　添加 jar 包步骤二

3. 常见错误

如果不添加外部包，会出现下列错误。

（1）错误 01：如图 7-15 所示。

```
java.sql.SQLException: Could not establish connection to hadoop11:10000/default: java.net.ConnectException: Connection refused: connect
    at org.apache.hadoop.hive.jdbc.HiveConnection.<init>(HiveConnection.java:117)
    at org.apache.hadoop.hive.jdbc.HiveDriver.connect(HiveDriver.java:106)
    at java.sql.DriverManager.getConnection(DriverManager.java:664)
    at java.sql.DriverManager.getConnection(DriverManager.java:247)
    at cn.orcale.com.bigdata.hive.HiveJdbcCli.getConn(HiveJdbcCli.java:153)
    at cn.orcale.com.bigdata.hive.HiveJdbcCli.main(HiveJdbcCli.java:33)
17/03/08 23:10:23 ERROR hive.HiveJdbcCli: Connection error!
java.sql.SQLException: Could not establish connection to hadoop11:10000/default: java.net.ConnectException: Connection refused: connect
    at org.apache.hadoop.hive.jdbc.HiveConnection.<init>(HiveConnection.java:117)
    at org.apache.hadoop.hive.jdbc.HiveDriver.connect(HiveDriver.java:106)
    at java.sql.DriverManager.getConnection(DriverManager.java:664)
    at java.sql.DriverManager.getConnection(DriverManager.java:247)
    at cn.orcale.com.bigdata.hive.HiveJdbcCli.getConn(HiveJdbcCli.java:153)
    at cn.orcale.com.bigdata.hive.HiveJdbcCli.main(HiveJdbcCli.java:33)
```

图 7-15　错误 01

（2）错误 02：如图 7-16 所示。

```
java.sql.SQLException: org.apache.thrift.transport.TTransportException
    at org.apache.hadoop.hive.jdbc.HiveStatement.executeQuery(HiveStatement.java:196)
    at org.apache.hadoop.hive.jdbc.HiveStatement.execute(HiveStatement.java:132)
    at org.apache.hadoop.hive.jdbc.HiveConnection.configureConnection(HiveConnection.java:133)
    at org.apache.hadoop.hive.jdbc.HiveConnection.<init>(HiveConnection.java:122)
    at org.apache.hadoop.hive.jdbc.HiveDriver.connect(HiveDriver.java:106)
    at java.sql.DriverManager.getConnection(DriverManager.java:664)
    at java.sql.DriverManager.getConnection(DriverManager.java:247)
    at cn.orcale.com.bigdata.hive.HiveJdbcCli.getConn(HiveJdbcCli.java:153)
    at cn.orcale.com.bigdata.hive.HiveJdbcCli.main(HiveJdbcCli.java:33)
17/03/08 23:19:38 ERROR hive.HiveJdbcCli: Connection error!
java.sql.SQLException: org.apache.thrift.transport.TTransportException
    at org.apache.hadoop.hive.jdbc.HiveStatement.executeQuery(HiveStatement.java:196)
    at org.apache.hadoop.hive.jdbc.HiveStatement.execute(HiveStatement.java:132)
    at org.apache.hadoop.hive.jdbc.HiveConnection.configureConnection(HiveConnection.java:133)
    at org.apache.hadoop.hive.jdbc.HiveConnection.<init>(HiveConnection.java:122)
    at org.apache.hadoop.hive.jdbc.HiveDriver.connect(HiveDriver.java:106)
```

图 7-16　错误 02

4. Java 连接 Hive 代码

Java 连接 Hive 代码的重点如下：

```
// before
private static String driverName =
"org.apache.hadoop.hive.jdbc.HiveDriver";
private static String url = "jdbc:hive://hadoop11:3306/hive";

// after
// private static String driverName =
"org.apache.hive.jdbc.HiveDriver";
//private static String url = "jdbc:hive2://tagtic-
master:10000/default";
```

```
package cn.orcale.com.bigdata.hive;

import java.sql.Connection;
import java.sql.DriverManager;
```

```java
import java.sql.ResultSet;
import java.sql.SQLException;
import java.sql.Statement;

import org.apache.log4j.Logger;

/***
 *
 * @author yuhui
 *
 */
public class HiveJdbcCli {

    // before
    private static String driverName = "org.apache.hadoop.hive.jdbc.HiveDriver";
    private static String url = "jdbc:hive://hadoop11:10000/hive";

    // after
    // private static String driverName = "org.apache.hive.jdbc.HiveDriver";
    // private static String url = "jdbc:hive2://tagtic-master:10000/default";
    private static String user = "root";
    private static String password = "123456";
    private static String sql = "";
    private static ResultSet res;
    private static final Logger log = Logger.getLogger(HiveJdbcCli.class);

    public static void main(String[] args) throws Exception {
        Connection conn = null;
        Statement stmt = null;
        try {
            conn = getConn();
            stmt = conn.createStatement();

            // 第一步:存在就先删除
            // String tableName = dropTable(stmt);

            // 第二步:不存在就创建
            // createTable(stmt, "hehe");

            // 第三步:查看创建的表
            showTables(stmt, "people");

            // 执行describe table操作
            // describeTables(stmt, tableName);

            // 执行load data into table操作
```

```java
            //    loadData(stmt, tableName);

            // 执行 select * query 操作
            //    selectData(stmt, tableName);

            // 执行 regular hive query 统计操作
            //    countData(stmt, tableName);

        } catch (ClassNotFoundException e) {
            e.printStackTrace();
            log.error(driverName + " not found!", e);
            System.exit(1);
        } catch (SQLException e) {
            e.printStackTrace();
            log.error("Connection error!", e);
            System.exit(1);
        } finally {
            try {
                if (conn != null) {
                    conn.close();
                    conn = null;
                }
                if (stmt != null) {
                    stmt.close();
                    stmt = null;
                }
            } catch (SQLException e) {
                e.printStackTrace();
            }
        }
    }

    private static void countData(Statement stmt, String tableName)
            throws SQLException {
        sql = "select count(1) from " + tableName;
        System.out.println("Running:" + sql);
        res = stmt.executeQuery(sql);
        System.out.println("执行"regular hive query"运行结果:");
        while (res.next()) {
            System.out.println("count ------>" + res.getString(1));
        }
    }

    private static void selectData(Statement stmt, String tableName)
            throws SQLException {
        sql = "select * from " + tableName;
        System.out.println("Running:" + sql);
        res = stmt.executeQuery(sql);
        System.out.println("执行 select * query 运行结果:");
        while (res.next()) {
```

```java
            System.out.println(res.getInt(1) + "\t" +
res.getString(2));
        }
    }

    private static void loadData(Statement stmt, String tableName)
            throws SQLException {
        String filepath = "/home/hadoop01/data";
        sql = "load data local inpath '" + filepath + "' into table "
                + tableName;
        System.out.println("Running:" + sql);
        res = stmt.executeQuery(sql);
    }

    private static void describeTables(Statement stmt, String tableName)
            throws SQLException {
        sql = "describe " + tableName;
        System.out.println("Running:" + sql);
        res = stmt.executeQuery(sql);
        System.out.println("执行 describe table 运行结果:");
        while (res.next()) {
            System.out.println(res.getString(1) + "\t" +
res.getString(2));
        }
    }

    private static void showTables(Statement stmt, String tableName)
            throws SQLException {
        sql = "show tables '" + tableName + "'";
        System.out.println("Running:" + sql);
        res = stmt.executeQuery(sql);
        System.out.println("执行 show tables 运行结果:");
        if (res.next()) {
            System.out.println(res.getString(1));
        }
    }

    private static void createTable(Statement stmt, String tableName)
            throws SQLException {
        sql = "create table "
                + tableName
                + " (key int, value string)  row format delimited fields terminated by '\t'";
        stmt.executeQuery(sql);
    }

    private static String dropTable(Statement stmt) throws SQLException {
        // 创建的表名
```

```
        String tableName = "testHive";
        sql = "drop table " + tableName;
        stmt.executeQuery(sql);
        return tableName;
    }

    private static Connection getConn() throws ClassNotFoundException,
            SQLException {
        Class.forName(driverName);
        Connection conn = DriverManager.getConnection(url, user, password);
        return conn;
    }

}
```

7.6 小结

  Hive 是基于 Hadoop 的一个数据仓库工具。本章讲解 Hive 组件：首先通过 Hive 简介和架构带领读者熟悉 Hive；接着讲解 Hive 的安装、Hive 的数据类型表结构、Hive 的分区、桶、数据倾斜；最后实战方面通过 Hive 客户端、Eclipse+Java 对 Hive 进行操作。

第 8 章

◀ Scala 实战 ▶

8.1 Scala 简介与安装

Scala 是一种有趣的语言。它一方面吸收继承了多种语言中的优秀特性，一方面又没有抛弃 Java 这个强大的平台，它运行在 Java 虚拟机（Java Virtual Machine）之上，轻松实现与丰富的 Java 类库互联互通。它既支持面向对象的编程方式，又支持函数式编程。它写出的程序像动态语言一样简洁，但事实上它确是严格意义上的静态语言。Scala 就像一位武林中的集大成者，将过去几十年计算机语言发展历史中的精粹集于一身，化繁为简，为程序员们提供了一种新的选择。

下载 Scala 进行安装，下载地址：http://www.scala-lang.org/download/。

文件名为 scala-2.11.8.msi，Scala 安装目录如图 8-1 所示。

图 8-1 Scala 安装目录

安装好后，Scala 的目录结构如图 8-2 所示。Scala 设置环境变量如图 8-3 所示。检验安装是否成功，可执行 scala -version 命令，Scala 测试如图 8-4 所示。

图 8-2　Scala 安装目录结构

图 8-3　Scala 设置环境变量

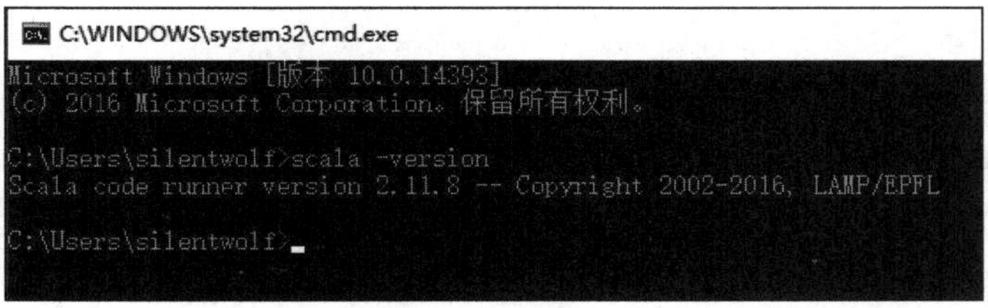

图 8-4　Scala 测试

8.2 IntelliJ IDEA 开发环境搭建

8.2.1 IntelliJ IDEA 简介

IntelliJ IDEA 2016.2 虽然不是最新版本，但却是目前工作中使用最多的 Java 集成开发环境，也被认为是当前 Java 开发效率最快的 IDE 工具。它整合了开发过程中实用的众多功能，包括智能代码助手、代码自动提示、重构、J2EE 支持、Ant、JUnit、CVS 整合、代码审查、创新的 GUI 设计等，几乎可以不用鼠标很方便地完成要做的任何事情，最大限度地加快开发的速度。

IntelliJ IDEA 2016.2 主要对前一版本的 Bug 进行了修复，还有一些可用性改进和性能提升。比如拥有更强大的文件搜索，Test 模式下的 UI 改进、强制返回、闭包 debug 等，功能强大而又实用，可以把 Java 开发人员从一些耗时的常规工作中解放出来，显著地提高了开发效率，绝对是当前最好的 Java 开发工具之一。

IntelliJ IDEA 2016.2 软件界面如图 8-5 所示。

图 8-5　IntelliJ IDEA 2016.2 软件界面

8.2.2 IntelliJ IDEA 安装

（1）下载安装包，解压缩，得到汉化包和软件原程序，文件名为 ideaIC-2016.2.exe。
（2）首先双击文件"ideaIC-2016.2.exe"，安装原版软件，界面如图 8-6 所示。

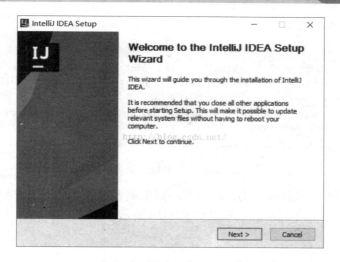

图 8-6　安装界面

（3）这里建议勾选三个选项，如图 8-7 所示。

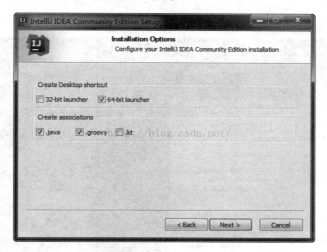

图 8-7　勾选三个选项

（4）成功安装后，双击桌面上生成的 IntelliJ IDEA Community Edition 2016.2(64)图标，打开软件，如图 8-8 所示。

图 8-8　IntelliJ IDEACommunity Edition 2016.2(64)图标

（5）这里可以暂时不设置，如图 8-9 所示。

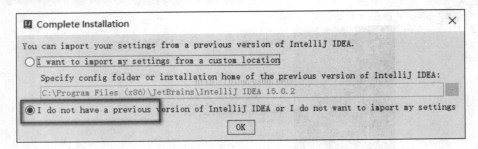

图 8-9　选择为安装过前版本的选项

（6）软件界面可以根据自己的喜好设置，这里笔者选择 IntelliJ IDEA 默认方式，再单击右下角按钮跳过。IntelliJ IDEA 背景设置如图 8-10 所示。

备注：

- IntelliJ 为默认背景模式，Darcula 为主题背景模式。
- skip All and Set Default 为设置默认（对于初用者）。
- Next : Default plugins 为可选配置（对于特殊需求者）。

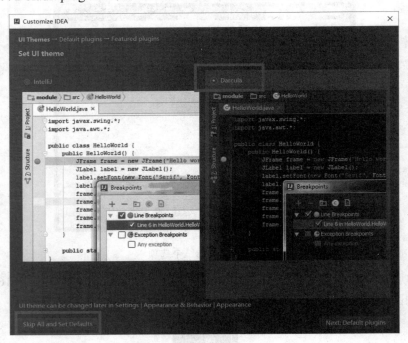

图 8-10　IntelliJ IDEA 背景设置

8.2.3　软件配置

1. 字体

在我们想修改字体的时候，Show only monospaced 项是选中状态，又不能够取消这个选项，只要单击 Save As，将 Default 保存一份就可以修改了。IntelliJ IDEA 字体设置如图 8-11 所示。

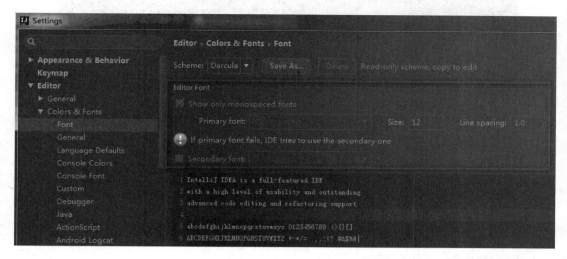

图 8-11　IntelliJ IDEA 字体设置

2. JAVA_SDK

IntelliJ IDEA 的 JAVA_SDK 设置如图 8-12 所示。

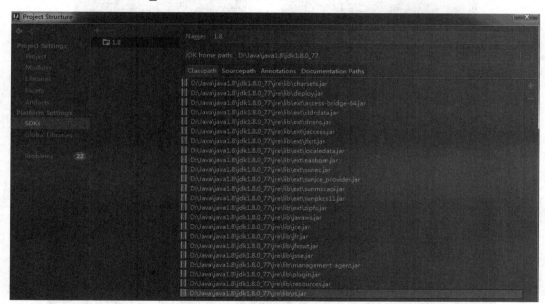

图 8-12　IntelliJ IDEA 的 JAVA_SDK 设置

3. 加载网络 Scala-SDK 插件

IntelliJ IDEA 插件有两种下载方式：在线下载与离线下载。加载网络 Scala-SDK 插件如图 8-13 所示。

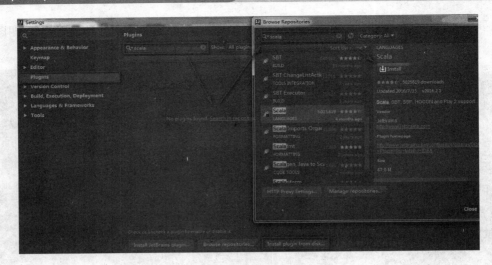

图 8-13　加载网络 Scala-SDK 插件

4. 加载本地 Scala-SDK 插件

加载本地 Scala-SDK 插件如图 8-14~图 8-16 所示。

 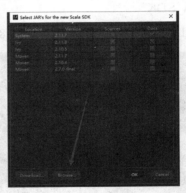

图 8-14　加载本地 Scala-SDK 插件一　　　图 8-15　加载本地 Scala-SDK 插件二

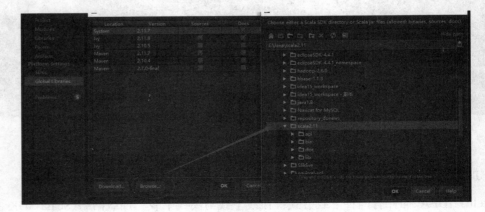

图 8-16　加载本地 Scala-SDK 插件三

5. 自动编译设置

自动编译设置如图 8-17 所示。

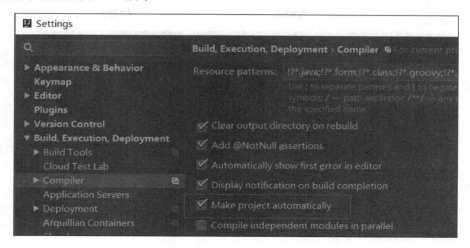

图 8-17　自动编译设置

6. Maven 配置

Maven 配置如图 8-18 所示。

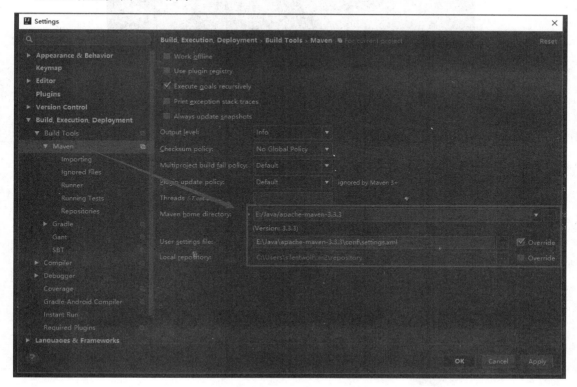

图 8-18　Maven 配置

7. 代码自动补齐

代码自动补齐如图 8-19 所示。

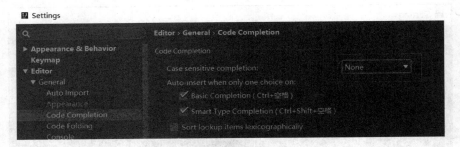

图 8-19　代码自动补齐

8. line_number 设置

line_number 设置如图 8-20 所示。

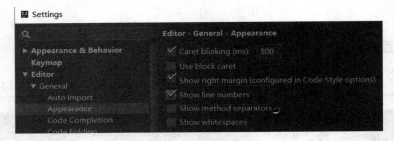

图 8-20　line_number 设置

9. 主题背景设置

主题背景设置如图 8-21 所示。

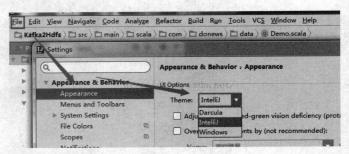

图 8-21　主题背景设置

8.3 IntelliJ IDEA 建立 Maven 项目

1. 建立 Maven 项目

建立 Maven 项目如图 8-22~图 8-26 所示。

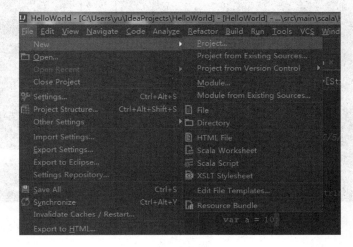

图 8-22　建立 Maven 项目一

图 8-23　建立 Maven 项目二

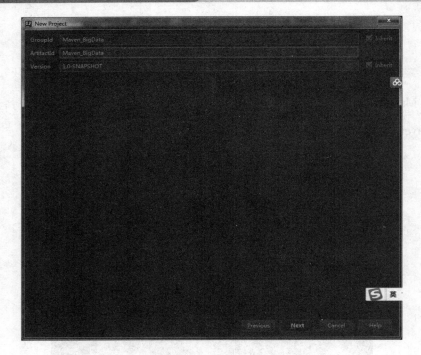

图 8-24　建立 Maven 项目三

图 8-25　建立 Maven 项目四

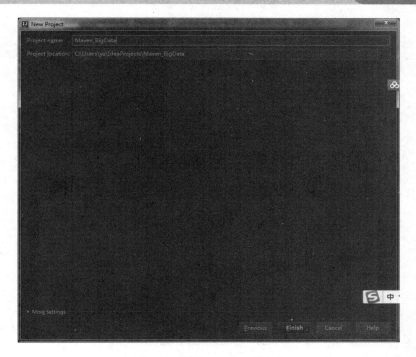

图 8-26　建立 Maven 项目五

单击 Finish 完成。

2. 建立目录和标记功能步骤

建立目录和标记功能步骤如图 8-27~图 8-29 所示。

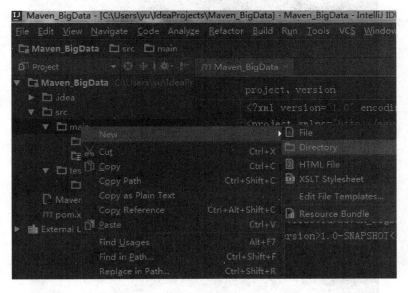

图 8-27　建立目录和标记功能一

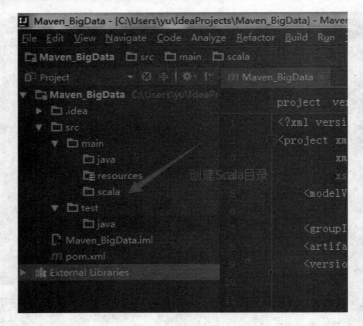

图 8-28　建立目录和标记功能二

图 8-29　建立目录和标记功能三

3. 建立包名和类名步骤

建立包名和类名步骤如图 8-30~图 8-35 所示。

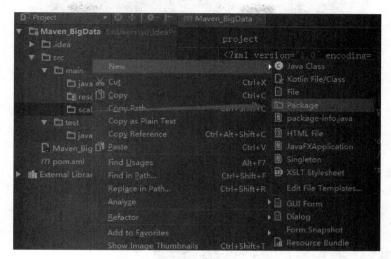

图 8-30　建立包名和类名一

图 8-31　建立包名和类名二

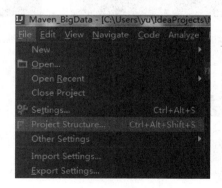

图 8-32　建立包名和类名三

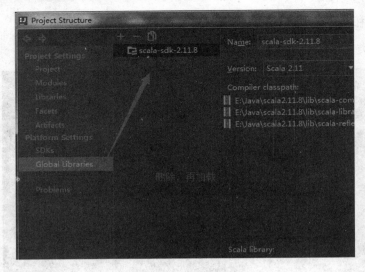

图 8-33 建立包名和类名四

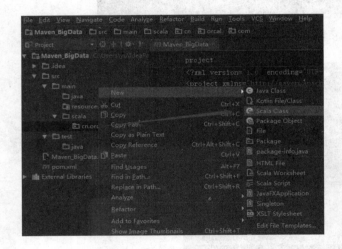

图 8-34 建立包名和类名五

图 8-35 建立包名和类名六

8.4 基础语法

1. 注释

```
// 单行注释开始于两个斜杠
/*
 *   多行注释，如你之前所见，看起来像这样
 */
```

2. 打印

// 打印并强制换行

```
println("Hello world!")
println(10)
```

// 没有强制换行的打印

```
print("Hello world")
```

3. 变量

// 通过 var 或者 val 来声明变量

// val 声明是不可变的，var 声明是可修改的。不可变性是好事。

```
val x = 10      // x 现在是 10
x = 20          // 错误：对 val 声明的变量重新赋值
var y = 10
y = 20          // y 现在是 20
```

4. 数据类型

Scala 与 Java 有着相同的数据类型，下面列出了 Scala 支持的数据类型如，图 8-36 所示。

数据类型	描述
Byte	8位有符号补码整数。数值区间为 -128 到 127
Short	16位有符号补码整数。数值区间为 -32768 到 32767
Int	32位有符号补码整数。数值区间为 -2147483648 到 2147483647
Long	64位有符号补码整数。数值区间为 -9223372036854775808 到 9223372036854775807
Float	32位IEEE754单精度浮点数
Double	64位IEEE754单精度浮点数
Char	16位无符号Unicode字符，区间值为 U+0000 到 U+FFFF
String	字符序列
Boolean	true或false
Unit	表示无值，和其他语言中void等同。用作不返回任何结果的方法的结果类型。Unit只有一个实例值，写成()。
Null	null 或空引用
Nothing	Nothing类型在Scala的类层级的最低端；它是任何其他类型的子类型。
Any	Any是所有其他类的超类
AnyRef	AnyRef类是Scala里所有引用类(reference class)的基类

图 8-36 Scala 支持的数据类型

Scala 数据类型设置样例：

```
val z: Int = 10
val a: Double = 1.0
```

注意从 Int 到 Double 的自动转型，结果是 10.0，不是 10：

```
val b: Double = 10.0
```

布尔值：

```
true
False
```

布尔操作：

```
!true           // false
!false          // true
true == false   // false
10 > 5          // true
```

5. 运算符

数学运算：

```
1 + 1    // 2
2 - 1    // 1
5 * 3    // 15
6 / 2    // 3
6 / 4    // 1
6.0 / 4  // 1.5
```

6. 字符串

```
"Scala strings are surrounded by double quotes"
'a' // Scala 的字符。
'不存在单引号字符串' <= 这会导致错误
```

String 有常见的 Java 字符串方法：

```
"hello world".length
"hello world".substring(2, 6)
"hello world".replace("C", "3")
```

也有一些额外的 Scala 方法：

```
"hello world".take(5)
"hello world".drop(5)
```

字符串改写，留意前缀 "s"：

```
val n = 45
s"We have $n apples" // => "We have 45 apples"
```

在要改写的字符串中使用表达式也是可以的：

```
val a = Array(11, 9, 6)
s"My second daughter is ${a(0) - a(2)} years old." // => "My second
s"We have double the amount of ${n / 2.0} in apples." // => "We hav
s"Power of 2: ${math.pow(2, 2)}" // => "Power of 2: 4"
```

添加 "f" 前缀对要改写的字符串进行格式化：

```
f"Power of 5: ${math.pow(5, 2)}%1.0f"   // "Power of 5: 25"
f"Square root of 122: ${math.sqrt(122)}%1.4f"   // "Square root of 12
```

未处理的字符串，忽略特殊字符：

```
raw"New line feed: \n. Carriage return: \r."   // => "New line feed:
```

一些字符需要转义，比如字符串中的双引号：

```
"They stood outside the \"Rose and Crown\""   // => "They stood outsi
```

三个双引号可以使字符串跨越多行，并包含引号：

```
val html = """<form id="daform">
<p>Press belo', Joe</p>
<input type="submit">
</form>"""
```

8.5 函数

函数是一组一起执行一个任务的语句。你可以把代码划分到不同的函数中。如何划分代码到不同的函数中是由你来决定的，但在逻辑上，划分通常是根据每个函数执行一个特定的任务来进行的。

Scala 有函数和方法，二者在语义上的区别很小。Scala 方法是类的一部分，而函数是一个对象，可以赋值给一个变量。换句话来说在类中定义的函数即是方法。

我们可以在任何地方定义函数，甚至可以在函数内定义函数（内嵌函数），更重要的一点是 Scala 函数名可以有以下特殊字符：+、++、~、&、-、--、\、/、: 等。

1. 函数声明

Scala 函数声明格式如下：

```
def functionName ([参数列表]) : [return type]
```

如果你不写等于号和方法主体，那么方法会被隐式声明为"抽象（abstract）"，于是包含它的类型也是一个抽象类型。

2. 函数定义

函数定义由一个 def 关键字开始，紧接着是可选的参数列表，一个冒号":" 和函数的返回类型，一个等于号"="，最后是函数的主体。

Scala 函数定义格式如下：

```
def functionName ([参数列表]) : [return type] = {
   function body
   return [expr]
}
```

以上代码中 return type 可以是任意合法的 Scala 数据类型。参数列表中的参数可以使用逗号分隔。

以下函数的功能是将两个传入的参数相加并求和：

```
object add{
  def addInt( a:Int, b:Int ) : Int = {
    var sum:Int = 0
    sum = a + b

    return sum
  }
}
```

如果函数没有返回值，可以返回为 Unit，这个类似于 Java 的 void，实例如下：

```
object Hello{
  def printMe( ) : Unit = {
    println("Hello, Scala!")
  }
}
```

3. 函数调用

Scala 提供了多种不同的函数调用方式。

以下是调用方法的标准格式：

```
functionName( 参数列表 )
```

如果函数使用了实例的对象来调用，我们可以使用类似 Java 的格式（使用 .号）：

```
[instance.]functionName( 参数列表 )
```

以上实例演示了定义与调用函数的实例：

```
object Test {
  def main(args: Array[String]) {
    println( "Returned Value : " + addInt(5,7) );
  }
  def addInt( a:Int, b:Int ) : Int = {
    var sum:Int = 0
    sum = a + b

    return sum
  }
}
```

执行以上代码，输出结果为：

```
Returned Value : 12
```

8.6 控制语句

1. 控制语句变量使用

Scala 对点和括号的要求非常宽松,注意其规则是不同的。这有助于写出读起来像英语的 DSL（领域特定语言）和 API（应用编程接口）。

```
1 to 5
val r = 1 to 5
r.foreach( println )
r foreach println
```

执行以上代码，输出结果为：

```
1,2,3,4,5,1
2
3
4
5
```

```
(5 to 1 by -1) foreach ( println )
```

执行以上代码，输出结果为：

```
5,4,3,2,1,
```

2. while 循环

运行一系列语句，如果条件为 true，会重复运行，直到条件变为 false。

```
var i = 0
while (i < 10) { println("i " + i); i+=1 }
```

执行以上代码，输出结果为：

```
i 0
i 1
i 2
i 3
i 4
i 5
i 6
i 7
i 8
i 9
```

3. do while 循环

类似 while 语句，区别在于判断循环条件之前，先执行一次循环的代码块。

```
var x = 0;
do {
    println(x + " is still less than 10");
    x += 1
} while (x < 10)
```

执行以上代码，输出结果为：

```
0 is still less than 10
1 is still less than 10
2 is still less than 10
3 is still less than 10
4 is still less than 10
5 is still less than 10
6 is still less than 10
7 is still less than 10
8 is still less than 10
9 is still less than 10
```

4. for 循环

for 循环允许编写一个执行指定次数的循环控制结构。

```
def main(args: Array[String]) {
    var a = 0;
    // for 循环
    for( a <- 1 to 10){
        println( "Value of a: " + a );
    }
}
```

执行以上代码，输出结果为：

```
value of a: 1
value of a: 2
value of a: 3
value of a: 4
value of a: 5
value of a: 6
value of a: 7
value of a: 8
value of a: 9
value of a: 10
```

5. 条件语句

Scala IF...ELSE 语句是通过一条或多条语句的执行结果（True 或者 False）来决定执行的代码块。

```
val x = 10
```

```
if (x == 1) println("yeah")
if (x == 10) println("yeah")
if (x == 11) println("yeah")
if (x == 11) println ("yeah") else println("nay")
println(if (x == 10) "yeah" else "nope")
val text = if (x == 10) "yeah" else "nope"
```

执行以上代码，输出结果为：

```
yeah
nay
yeah
```

6. break 语句

Scala 语言中默认是没有 break 语句，但是在 Scala 2.8 版本后可以使用另外一种方式来实现 break 语句。当在循环中使用 break 语句，在执行到该语句时，就会中断循环并执行循环体之后的代码块。

Scala 中 break 的语法有点不大一样，格式如下：

```
// 导入以下包
import scala.util.control._

// 创建 Breaks 对象
val loop = new Breaks;

// 在 breakable 中循环
loop.breakable{
    // 循环
    for(...){
       ....
       // 循环中断
       loop.break;
    }
}
```

实例：

```
import scala.util.control._

object Test {
   def main(args: Array[String]) {
      var a = 0;
      val numList = List(1,2,3,4,5,6,7,8,9,10);

      val loop = new Breaks;
      loop.breakable {
        for( a <- numList){
           println( "Value of a: " + a );
           if( a == 4 ){
              loop.break;
```

```
        }
      }
    }
    println( "After the loop" );
  }
}
```

执行以上代码输出结果为:

```
Value of a: 1
Value of a: 2
Value of a: 3
Value of a: 4
After the loop
```

8.7 函数式编程

1. Array(数组)

Scala 数组声明的语法格式:

```
var z:Array[String] = new Array[String](3)
或
var z = new Array[String](3)
```

数组的元素类型和数组的大小都是确定的,所以当处理数组元素时,我们通常使用基本的 for 循环。

以下实例演示了数组的创建、初始化等处理过程:

```
object Test {
  def main(args: Array[String]) {
    var myList = Array(1.9, 2.9, 3.4, 3.5)

    // 输出所有数组元素
    for ( x <- myList ) {
      println( x )
    }

    // 计算数组所有元素的总和
    var total = 0.0;
    for ( i <- 0 to (myList.length - 1)) {
      total += myList(i);
    }
    println("总和为 " + total);

    // 查找数组中的最大元素
```

```
      var max = myList(0);
      for ( i <- 1 to (myList.length - 1) ) {
         if (myList(i) > max) max = myList(i);
      }
      println("最大值为 " + max);

   }
}
```

执行以上代码,输出结果为:

```
1.9
2.9
3.4
3.5
总和为 11.7
最大值为 3.5
```

2. List(列表)

List 的特征是其元素以线性方式存储,集合中可以存放重复对象。

以下列出了多种类型的列表:

```
// 字符串列表
val site: List[String] = List("北京小辉的博客", "Google", "Baidu")

// 整型列表
val nums: List[Int] = List(1, 2, 3, 4)

// 空列表
val empty: List[Nothing] = List()

// 二维列表
val dim: List[List[Int]] =
   List(
      List(1, 0, 0),
      List(0, 1, 0),
      List(0, 0, 1)
   )
```

对于 Scala 列表的任何操作都可以使用这三个基本操作来表达,实例如下:

```
object Test {
   def main(args: Array[String]) {
      val site = "北京小辉的博客" :: ("Google" :: ("Baidu" :: Nil))
      val nums = Nil

      println( "第一网站是 : " + site.head )
      println( "最后一个网站是 : " + site.tail )
      println( "查看列表 site 是否为空 : " + site.isEmpty )
      println( "查看 nums 是否为空 : " + nums.isEmpty )
   }
```

执行以上代码，输出结果为：

```
第一网站是 : 北京小辉的博客
最后一个网站是 : List(Google, Baidu)
查看列表 site 是否为空 : false
查看 nums 是否为空 : true
```

3. Set（集合）

Set 是最简单的一种集合。集合中的对象不按特定的方式排序，并且没有重复对象。

对于 Scala 集合的任何操作都可以使用这三个基本操作来表达，实例如下：

```
object Test {
  def main(args: Array[String]) {
    val site = Set("北京小辉的博客", "Google", "Baidu")
    val nums: Set[Int] = Set()

    println( "第一网站是 : " + site.head )
    println( "最后一个网站是 : " + site.tail )
    println( "查看列表 site 是否为空 : " + site.isEmpty )
    println( "查看 nums 是否为空 : " + nums.isEmpty )
  }
}
```

执行以上代码，输出结果为：

```
第一网站是 : 北京小辉的博客
最后一个网站是 : Set(Google, Baidu)
查看列表 site 是否为空 : false
查看 nums 是否为空 : true
```

4. Map（映射）

Map 是一种把键对象和值对象映射的集合，它的每一个元素都包含一对键对象和值对象。

以下实例演示了以上三个方法的基本应用：

```
object Test {
  def main(args: Array[String]) {
    val colors = Map("red" -> "#FF0000",
                     "azure" -> "#F0FFFF",
                     "peru" -> "#CD853F")

    val nums: Map[Int, Int] = Map()

    println( "colors 中的键为 : " + colors.keys )
    println( "colors 中的值为 : " + colors.values )
    println( "检测 colors 是否为空 : " + colors.isEmpty )
    println( "检测 nums 是否为空 : " + nums.isEmpty )
  }
}
```

```
}
```

执行以上代码，输出结果为：

```
colors 中的键为 : Set(red, azure, peru)
colors 中的值为 : MapLike(#FF0000, #F0FFFF, #CD853F)
检测 colors 是否为空 : false
检测 nums 是否为空 : true
```

5. 元组

元组是不同类型的值的集合。

与列表一样，元组也是不可变的，但与列表不同的是元组可以包含不同类型的元素。

元组的值是通过将单个的值包含在圆括号中构成的。例如：

```
val t = (1, 3.14, "Fred")
```

以上实例在元组中定义了三个元素，对应的类型分别为[Int, Double, java.lang.String]。

此外也可以使用以下方式来定义：

```
val t = new Tuple3(1, 3.14, "Fred")
```

可以使用 t._1 访问第一个元素， t._2 访问第二个元素等，如下所示：

```
object Test {
   def main(args: Array[String]) {
      val t = (4,3,2,1)

      val sum = t._1 + t._2 + t._3 + t._4

      println( "元素之和为: "  + sum )
   }
}
```

执行以上代码，输出结果为：

```
元素之和为: 10
```

6. Option

Option[T] 表示有可能包含值的容器，也可能不包含值。Scala Iterator（迭代器）不是一个容器，更确切地说是逐一访问容器内元素的方法。Scala Option（选项）类型用来表示一个值是可选的（有值或无值）。

Option[T] 是一个类型为 T 的可选值的容器：如果值存在，Option[T] 就是一个 Some[T]；如果不存在，Option[T] 就是对象 None 。

接下来我们来看一段代码：

```
// 虽然 Scala 可以不定义变量的类型,不过为了清楚些,作者还是
// 把他显示的定义上了

val myMap: Map[String, String] = Map("key1" -> "value")
val value1: Option[String] = myMap.get("key1")
```

```
val value2: Option[String] = myMap.get("key2")

println(value1) // Some("value1")
println(value2) // None
```

在上面的代码中，myMap 是一个 Key 的类型是 String，Value 的类型是 String 的 hash map，但不一样的是他的 get() 返回的是一个叫 Option[String] 的类别。

Scala 使用 Option[String] 来告诉你："我会想办法回传一个 String，但也可能没有 String 给你"。

myMap 里并没有 key2 这笔数据，get() 方法返回 None。

Option 有两个子类别，一个是 Some，一个是 None，当它回传 Some 的时候，代表这个函数成功地给了你一个 String，而你可以透过 get() 这个函数拿到那个 String；如果它返回的是 None，则代表没有字符串可以给你。

另一个实例：

```
object Test {
  def main(args: Array[String]) {
    val sites = Map("余辉" -> "北京小辉的博客", "google" -> "www.google.com")

    println("sites.get( \"余辉\" ) : " + sites.get( "余辉" )) // Some(www.runoob.com)

    println("sites.get( \"baidu\" ) : " + sites.get( "baidu" )) // None
  }
}
```

执行以上代码，输出结果为：

```
sites.get( "runoob" ) : Some(北京小辉的博客)
sites.get( "baidu" ) : None
```

也可以通过模式匹配来输出匹配值，实例如下：

```
object Test {
  def main(args: Array[String]) {
    val sites = Map("余辉" -> "北京小辉的博客", "google" -> "www.google.com")

    println("show(sites.get( \"余辉\")) : " +
                      show(sites.get( "余辉")) )
    println("show(sites.get( \"baidu\")) : " +
                      show(sites.get( "baidu")) )
  }

  def show(x: Option[String]) = x match {
    case Some(s) => s
    case None => "?"
  }
}
```

```
}
```

执行以上代码,输出结果为:

```
show(sites.get( "余辉")) : 北京小辉的博客
show(sites.get( "baidu")) : ?
```

8.8 模式匹配

Scala 提供了强大的模式匹配机制,应用也非常广泛。

一个模式匹配包含了一系列备选项,每个都开始于关键字 case。每个备选项都包含了一个模式及一到多个表达式。箭头符号 => 隔开了模式和表达式。

以下是一个简单的整型值模式匹配实例:

```
object Test {
  def main(args: Array[String]) {
    println(matchTest(3))

  }
  def matchTest(x: Int): String = x match {
    case 1 => "one"
    case 2 => "two"
    case _ => "many"
  }
}
```

执行以上代码,输出结果为:

```
many
```

match 对应 Java 里的 switch,但是写在选择器表达式之后。即:选择器 match {备选项}。

match 表达式通过以代码编写的先后次序尝试每个模式来完成计算,只要发现有一个匹配的 case,剩下的 case 不会继续匹配。

接下来我们来看一个不同数据类型的模式匹配:

```
object Test {
  def main(args: Array[String]) {
    println(matchTest("two"))
    println(matchTest("test"))
    println(matchTest(1))
    println(matchTest(6))

  }
  def matchTest(x: Any): Any = x match {
    case 1 => "one"
    case "two" => 2
    case y: Int => "scala.Int"
```

```
        case _ => "many"
    }
}
```

执行以上代码，输出结果为：

```
2
many
one
scala.Int
```

实例中第 1 个 case 对应整型数值 1，第 2 个 case 对应字符串值 two，第 3 个 case 对应类型模式，用于判断传入的值是否为整型，相比使用 isInstanceOf 来判断类型，使用模式匹配更好。第 4 个 case 表示默认的全匹配备选项，即没有找到其他匹配时的匹配项，类似 switch 中的 default。

使用样例类

使用了 case 关键字的类定义就是样例类（case classes），样例类是一种特殊的类，经过优化后用于模式匹配。

以下是样例类的简单实例：

```
object Test {
  def main(args: Array[String]) {
   val alice = new Person("Alice", 25)
   val bob = new Person("Bob", 32)
   val charlie = new Person("Charlie", 32)

   for (person <- List(alice, bob, charlie)) {
   person match {
        case Person("Alice", 25) => println("Hi Alice!")
        case Person("Bob", 32) => println("Hi Bob!")
        case Person(name, age) =>
           println("Age: " + age + " year, name: " + name + "?")
      }
    }
  }
  // 样例类
  case class Person(name: String, age: Int)
}
```

执行以上代码，输出结果为：

```
Hi Alice!
Hi Bob!
Age: 32 year, name: Charlie?
```

8.9 类和对象

1. 类的定义

类是对象的抽象，而对象是类的具体实例。类是抽象的，不占用内存，而对象是具体的，占用存储空间。类是用于创建对象的蓝图，它是一个定义包括在特定类型的对象中的方法和变量的软件模板。

我们可以使用 new 关键字来创建类的对象，实例如下：

```
class Point(xc: Int, yc: Int) {
   var x: Int = xc
   var y: Int = yc

   def move(dx: Int, dy: Int) {
      x = x + dx
      y = y + dy
      println ("x 的坐标点: " + x);
      println ("y 的坐标点: " + y);
   }
}
```

Scala 中的类不声明为 public，一个 Scala 源文件中可以有多个类。

以上实例的类定义了两个变量 x 和 y，一个方法：move，方法没有返回值。

Scala 的类定义可以有参数，称为类参数，如上面的 xc、yc，类参数在整个类中都可以访问。

接着可以使用 new 来实例化类，并访问类中的方法和变量：

```
import java.io._

class Point(xc: Int, yc: Int) {
   var x: Int = xc
   var y: Int = yc

   def move(dx: Int, dy: Int) {
      x = x + dx
      y = y + dy
      println ("x 的坐标点: " + x);
      println ("y 的坐标点: " + y);
   }
}

object Test {
   def main(args: Array[String]) {
      val pt = new Point(10, 20);

      // 移到一个新的位置
      pt.move(10, 10);
```

```
    }
}
```

执行以上代码,输出结果为:

```
x 的坐标点: 20
y 的坐标点: 30
```

2. 继承

Scala 继承一个基类跟 Java 很相似,但需要注意以下几点:

(1) 重写一个非抽象方法必须使用 override 修饰符。
(2) 只有主构造函数才可以往基类的构造函数里写参数。
(3) 在子类中重写超类的抽象方法时,不需要使用 override 关键字。

接下来让我们来看个实例:

```
class Point(xc: Int, yc: Int) {
   var x: Int = xc
   var y: Int = yc

   def move(dx: Int, dy: Int) {
      x = x + dx
      y = y + dy
      println ("x 的坐标点: " + x);
      println ("y 的坐标点: " + y);
   }
}
class Location(override val xc: Int, override val yc: Int,
   val zc :Int) extends Point(xc, yc){
   var z: Int = zc

   def move(dx: Int, dy: Int, dz: Int) {
      x = x + dx
      y = y + dy
      z = z + dz
      println ("x 的坐标点 : " + x);
      println ("y 的坐标点 : " + y);
      println ("z 的坐标点 : " + z);
   }
}
```

Scala 使用 extends 关键字来继承一个类。实例中 Location 类继承了 Point 类。Point 称为父类(基类),Location 称为子类。

override val xc 为重写了父类的字段。

继承会继承父类的所有属性和方法,Scala 只允许继承一个父类。

实例如下:

```
import java.io.
```

```scala
class Point(val xc: Int, val yc: Int) {
   var x: Int = xc
   var y: Int = yc
   def move(dx: Int, dy: Int) {
      x = x + dx
      y = y + dy
      println ("x 的坐标点 : " + x);
      println ("y 的坐标点 : " + y);
   }
}

class Location(override val xc: Int, override val yc: Int,
   val zc :Int) extends Point(xc, yc){
   var z: Int = zc

   def move(dx: Int, dy: Int, dz: Int) {
      x = x + dx
      y = y + dy
      z = z + dz
      println ("x 的坐标点 : " + x);
      println ("y 的坐标点 : " + y);
      println ("z 的坐标点 : " + z);
   }
}

object Test {
   def main(args: Array[String]) {
      val loc = new Location(10, 20, 15);

      // 移到一个新的位置
      loc.move(10, 10, 5);
   }
}
```

执行以上代码，输出结果为：

```
x 的坐标点 : 20
y 的坐标点 : 30
z 的坐标点 : 20
```

Scala 重写一个非抽象方法，必须用 override 修饰符。

```scala
class Person {
  var name = ""
  override def toString = getClass.getName + "[name=" + name + "]"
}

class Employee extends Person {
  var salary = 0.0
  override def toString = super.toString + "[salary=" + salary + "]"
}
```

```
object Test extends App {
  val fred = new Employee
  fred.name = "Fred"
  fred.salary = 50000
  println(fred)
}
```

执行以上代码，输出结果为：

```
Employee[name=Fred][salary=50000.0]
```

8.10 Scala 异常处理

Scala 的异常处理与其他语言比如 Java 类似。
Scala 的方法可以通过抛出异常方法的方式来终止相关代码的运行，不必通过返回值。

1. 抛出异常

Scala 抛出异常的方法和 Java 一样，使用 throw 方法。例如，抛出一个新的参数异常：

```
throw new IllegalArgumentException
```

2. 捕获异常

异常捕捉的机制与其他语言的处理方法一样，如果有异常发生，catch 子句是按次序捕捉的。因此，在 catch 子句中，越具体的异常越要靠前，越普遍的异常越靠后。如果抛出的异常不在 catch 子句中，该异常则无法处理，会被升级到调用者处。

捕捉异常的 catch 子句，语法与其他语言中不太一样。在 Scala 里，借用了模式匹配的思想来做异常的匹配，因此，在 catch 的代码里，是一系列 case 字句，如下例所示：

```
import java.io.FileReader
import java.io.FileNotFoundException
import java.io.IOException

object Test {
  def main(args: Array[String]) {
    try {
      val f = new FileReader("input.txt")
    } catch {
      case ex: FileNotFoundException =>{
        println("Missing file exception")
      }
      case ex: IOException => {
        println("IO Exception")
      }
    }
  }
}
```

```
    }
}
```

执行以上代码，输出结果为：

```
Missing file exception
```

catch 语句里的内容跟 match 里的 case 是完全一样的。由于异常捕捉是按次序，如果把最普遍的异常 Throwable 写在最前面，则在它后面的 case 都捕捉不到，因此需要将它写在最后面。

3. finally 语句

finally 语句用于执行不管是正常处理还是有异常发生时都需要执行的步骤，实例如下：

```
import java.io.FileReader
import java.io.FileNotFoundException
import java.io.IOException

object Test {
   def main(args: Array[String]) {
      try {
         val f = new FileReader("input.txt")
      } catch {
         case ex: FileNotFoundException => {
            println("Missing file exception")
         }
         case ex: IOException => {
            println("IO Exception")
         }
      } finally {
         println("Exiting finally...")
      }
   }
}
```

执行以上代码，输出结果为：

```
Missing file exception
Exiting finally...
```

8.11 Trait（特征）

Scala Trait（特征）相当于 Java 的接口，实际上它比接口的功能还要强大。
与接口不同的是，它还可以定义属性和方法的实现。
一般情况下 Scala 的类只能够继承单一父类，但是如果是 Trait 的话就可以继承多个，从结果来看就是实现了多重继承。
Trait 定义的方式与类类似，但它使用的关键字是 trait，如下所示：

```
trait Equal {
  def isEqual(x: Any): Boolean
  def isNotEqual(x: Any): Boolean = !isEqual(x)
}
```

以上 Trait 由两个方法组成：isEqual 和 isNotEqual。isEqual 方法没有定义方法的实现，isNotEqual 定义了方法的实现。子类继承特征可以实现未被实现的方法。所以其实 Scala Trait 更像 Java 的抽象类。

以下演示了特征的完整实例：

```
trait Equal {
  def isEqual(x: Any): Boolean
  def isNotEqual(x: Any): Boolean = !isEqual(x)
}

class Point(xc: Int, yc: Int) extends Equal {
  var x: Int = xc
  var y: Int = yc
  def isEqual(obj: Any) =
    obj.isInstanceOf[Point] &&
    obj.asInstanceOf[Point].x == x
}

object Test {
  def main(args: Array[String]) {
    val p1 = new Point(2, 3)
    val p2 = new Point(2, 4)
    val p3 = new Point(3, 3)

    println(p1.isNotEqual(p2))
    println(p1.isNotEqual(p3))
    println(p1.isNotEqual(2))
  }
}
```

执行以上代码，输出结果为：

```
false
true
true
```

8.12 Scala 文件 I/O

1. IO 介绍

Scala 进行文件写操作，直接用的都是 Java 中的 I/O 类（java.io.File）：

```
import java.io._
```

```
object Test {
  def main(args: Array[String]) {
    val writer = new PrintWriter(new File("test.txt" ))

    writer.write("北京小辉的博客 http://blog.csdn.net/silentwolfyh")
    writer.close()
  }
}
```

执行以上代码，会在你的当前目录下生产一个 test.txt 文件，文件内容为"北京小辉的博客 http://blog.csdn.net/silentwolfyh"。

2. 从屏幕上读取用户输入

有时候我们需要接收用户在屏幕输入的指令来处理程序，实例如下：

```
object Test {
  def main(args: Array[String]) {
    print("请输入北京小辉的博客： " )
    val line = Console.readLine

    println("谢谢，你输入的是： " + line)
  }
}
```

执行以上代码，屏幕上会显示如下信息：

```
请输入北京小辉的博客： http://blog.csdn.net/silentwolfyh
谢谢，你输入的是： http://blog.csdn.net/silentwolfyh
```

3. 从文件上读取内容

从文件读取内容非常简单。我们可以使用 Scala 的 Source 类及伴生对象来读取文件。以下实例演示了从 "test.txt"（之前已创建过）文件中读取内容：

```
import scala.io.Source

object Test {
  def main(args: Array[String]) {
    println("文件内容为:" )

    Source.fromFile("test.txt" ).foreach{
      print
    }
  }
}
```

执行以上代码，输出结果为：

```
文件内容为:
北京小辉的博客 http://blog.csdn.net/silentwolfyh
```

8.13 作业

8.13.1 九九乘法表

本次作业，掌握：变量、循环、运算符、数据类型。

```
/**
  * Created by yuhui on 2017/6/4.
  */
object chengfabiao {
  def main(args: Array[String]) {
    var i: Int = 1
    while (i <= 9) {
      {
        var j: Int = 1
        while (j <= i) {
          {
            System.out.print(i + "*" + j + "=" + (i * j) + "\t")
          }
          {
            j += 1; j - 1
          }
        }
        System.out.print("\n")
      }
      {
        i += 1; i - 1
      }
    }
  }
}
```

九九乘法表执行结果如图 8-37 所示。

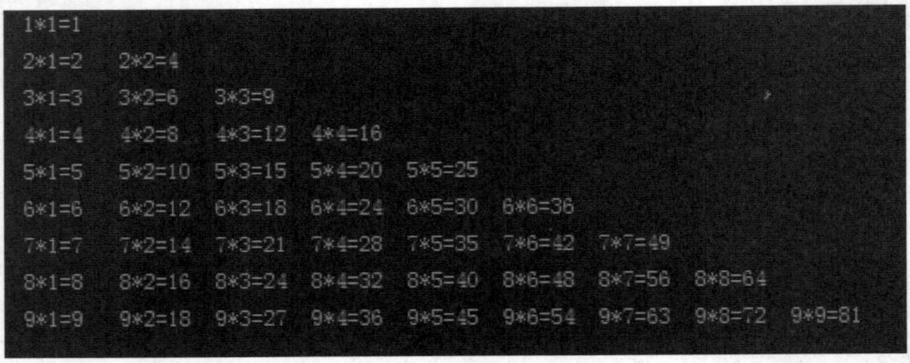

```
1*1=1
2*1=2	2*2=4
3*1=3	3*2=6	3*3=9
4*1=4	4*2=8	4*3=12	4*4=16
5*1=5	5*2=10	5*3=15	5*4=20	5*5=25
6*1=6	6*2=12	6*3=18	6*4=24	6*5=30	6*6=36
7*1=7	7*2=14	7*3=21	7*4=28	7*5=35	7*6=42	7*7=49
8*1=8	8*2=16	8*3=24	8*4=32	8*5=40	8*6=48	8*7=56	8*8=64
9*1=9	9*2=18	9*3=27	9*4=36	9*5=45	9*6=54	9*7=63	9*8=72	9*9=81
```

图 8-37　九九乘法表执行结果

8.13.2 冒泡排序

本次作业，掌握：数组、变量、循环、判断、数据类型。

```scala
/**
 * Created by yuhui on 2017/6/4.
 */
object maopao {
  def main(args: Array[String]) {
    val score: Array[Int] = Array(67, 55, 75, 87, 89, 70, 99, 10)
    var i: Int = 0
    while (i < score.length - 1) {
      {
        //最多做n-1趟排序
        var j: Int = 0
        while (j < score.length - i - 1) {
          {
            //对当前无序区间score[0......length-i-1]进行排序(j的范围很关键,这个范围是在逐步缩小的)
            if (score(j) < score(j + 1)) {
              //把小的值交换到后面
              val temp: Int = score(j)
              score(j) = score(j + 1)
              score(j + 1) = temp
            }
          }
          {
            j += 1; j - 1
          }
        }
        System.out.print("第" + (i + 1) + "次排序结果：")
        var a: Int = 0
        while (a < score.length) {
          {
            System.out.print(score(a) + "\t")
          }
          {
            a += 1; a - 1
          }
        }
        System.out.println("")
      }
      {
        i += 1; i - 1
      }
    }
    System.out.print("最终排序结果：")
    var a: Int = 0
    while (a < score.length) {
      {
        System.out.print(score(a) + "\t")
```

```
        }
        {
          a += 1; a - 1
        }
      }
    }
}
```

冒泡排序执行结果如图 8-38 所示。

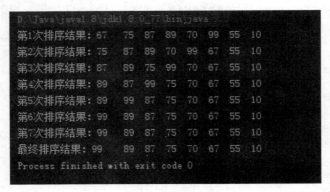

图 8-38 冒泡排序执行结果

8.13.3 设计模式 Command

本次作业，掌握：面向对象、模式匹配、继承、特征。设计模式 Command 代码结构如图 8-39 所示。

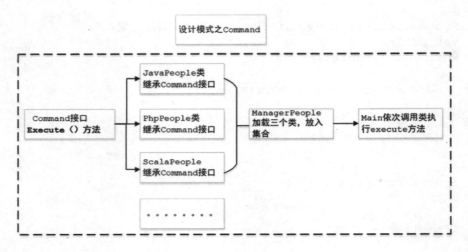

图 8-39 设计模式 Command 代码结构

```
/**
 * Created by yuhui on 2017/6/4.
 */
```

```
trait moshi {
  def work()
}

object JavaPeople extends moshi{
  override def work(): Unit ={
    println("我是一个Java工程师")
  }
}

object PhpPeople extends moshi{
  override def work(): Unit ={
    println("我是一个PhpPeople工程师")
  }
}

object ScalaPeople extends moshi{
  override def work(): Unit ={
    println("我是一个Spark工程师")
  }
}

object ManagerPeople{
  def main(args: Array[String]): Unit = {

    val work01 = "java"
    matchTest(work01)
    val work02 = "php"
    matchTest(work02)
    val work03 = "spark"
    matchTest(work03)
    val work04 = "hehe"
    matchTest(work04)
  }

  def matchTest(x: String)= x match {
    case "java" => JavaPeople.work()
    case "php" => PhpPeople.work()
    case "spark" => ScalaPeople.work()
    case _ => println("I don't know your job")
  }
}
```

设计模式Command执行结果如图8-40所示。

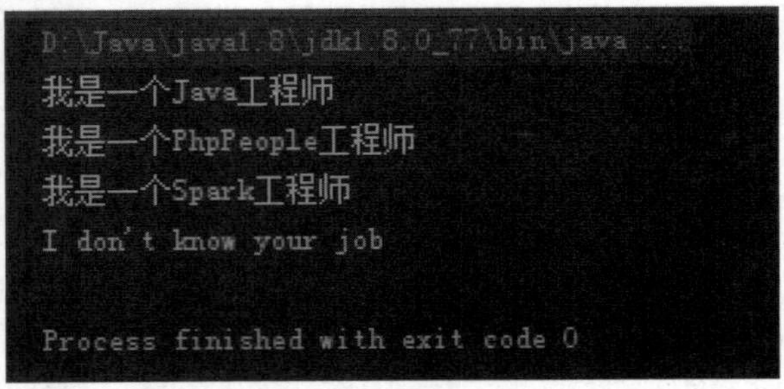

图 8-40 设计模式 Command 执行结果

8.13.4 集合对称判断

本次作业，掌握：集合、判断、循环。

有两个集合如下：

- 集合 (1,2,3,2,1)
- 集合 (1,2,3,3,2,1)

需求：判断这两个链表是否对称。对称返回 true，不对称返回 false。

```
/**
 * Created by yuhui on 2017/6/4.
 */
object duichen {
  def main(args: Array[String]) {
    val list : util.LinkedList[String]= new util.LinkedList[String]()
    list.add("1")
    list.add("2")
    list.add("3")
    list.add("4")
    list.add("3")
    list.add("2")
    list.add("1")

    if (list.size % 2 == 0) {
      var i: Int = 0
      while (i <= list.size / 2 - 1) {
        {
          if (i != (list.size - 1) / 2 + 1) if (list.get(i) .equals(list.get(list.size - 1 - i)) ) {

            System.out.println(list.get(i) + "----" + list.get(list.size - 1 - i))
            if (i == list.size / 2 - 1) System.out.println("true")
          }
          else {
            System.out.println(list.get(i) + "----" +
```

```
list.get(list.size - 1 - i) + "出现错误")
              System.out.println("flase")
              return "false";//todo: break is not supported
            }
          }
          {
            i += 1; i - 1
          }
        }
      }
      else {
        var i: Int = 0
        while (i <= (list.size - 1) / 2 - 1) {
          {
            if (i != (list.size - 1) / 2 + 1) if (list.get(i) eq
list.get(list.size - 1 - i)) {
              System.out.println(list.get(i) + "----" +
list.get(list.size - 1 - i))
              if (i == list.size / 2 - 1) System.out.println("true")
            }
            else {
              System.out.println(list.get(i) + "----" +
list.get(list.size - 1 - i) + "出现错误")
              System.out.println("flase")
              return "false";
            }
          }
          {
            i += 1; i - 1
          }
        }
      }
    }
}
```

链表对称判断执行结果如图 8-41 所示。

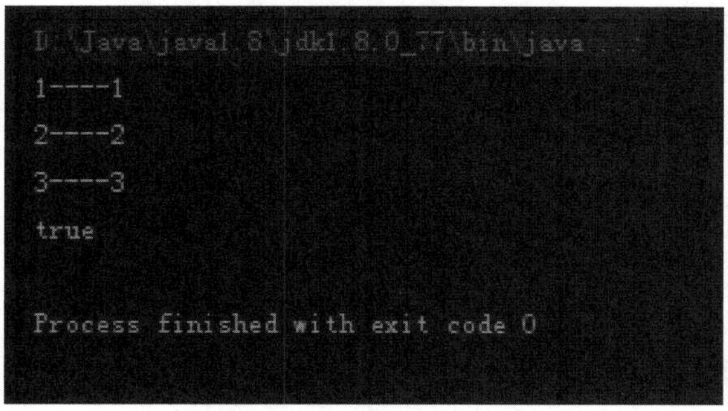

图 8-41　链表对称判断执行结果

8.13.5 综合题

本次作业掌握：IO、模式匹配、正则、HTTP 请求、导入类。

本次作业需求：

（1）读取数据文件。

（2）提取 IP 字段。

（3）使用正则和模式匹配，判断是否是 IP。

（4）如果是 IP，则使用 get 方法调用淘宝 IP 接口。

（5）如果不是 IP，则不处理。

（6）将结果数据存储到本地。

原始日志数据：

```
1436992033 123.150.156.4 d77dbbbd035355347be6c52305d89281
   c007235244ad9d8edd3d8f082ff9ed3d   si2.mfniu.com 1920   1080   32
   zh-CN
   http://si2.mfniu.com/HTML_Content_Cache/stock_indicator/1/601727.
html   http://si2.mfniu.com/    Mozilla/5.0 (Windows NT 6.1)
AppleWebKit/537.36 (KHTML, like Gecko) Chrome/31.0.1650.63
Safari/537.36 huatuozhenguseo  _trackPageview       si2.mfniu.com
   10119
1436993140 110.84.190.37 f0171969068dc342029323ca9bc42c10
   8b631869a0f53c227600859deb0bb777   si.mfniu.com  1920   1080   32
      http://si.mfniu.com/Default.aspx
   http://si.mfniu.com/Hot.aspx    Mozilla/5.0 (compatible; MSIE
9.0; Windows NT 6.1; WOW64; Trident/5.0)  huatuozhengu
    trackPageview       si.mfniu.com  10119
```

原始日志图片如图 8-42 所示。

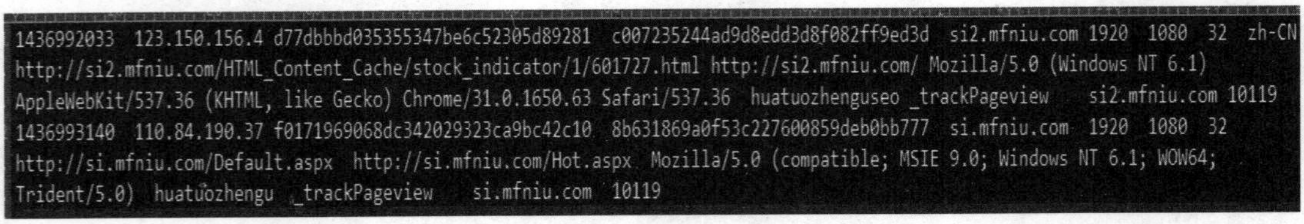

1436992033 123.150.156.4 d77dbbbd035355347be6c52305d89281 c007235244ad9d8edd3d8f082ff9ed3d si2.mfniu.com 1920 1080 32 zh-CN
http://si2.mfniu.com/HTML_Content_Cache/stock_indicator/1/601727.html http://si2.mfniu.com/ Mozilla/5.0 (Windows NT 6.1)
AppleWebKit/537.36 (KHTML, like Gecko) Chrome/31.0.1650.63 Safari/537.36 huatuozhenguseo _trackPageview si2.mfniu.com 10119
1436993140 110.84.190.37 f0171969068dc342029323ca9bc42c10 8b631869a0f53c227600859deb0bb777 si.mfniu.com 1920 1080 32
http://si.mfniu.com/Default.aspx http://si.mfniu.com/Hot.aspx Mozilla/5.0 (compatible; MSIE 9.0; Windows NT 6.1; WOW64;
Trident/5.0) huatuozhengu _trackPageview si.mfniu.com 10119

图 8-42　原始日志

```
package com.ou.cn.test

import java.io.{File, PrintWriter}

import scala.io.Source
/**
  * Created by yuhui on 2017/6/4.
  */
object duxie {
```

```scala
def main(args: Array[String]): Unit = {
  val writer = new PrintWriter(new File("D:\\test.txt" ))
  val r=Source.fromFile("D:\\NginxData.txt" ).getLines.toList

  for (i <- 0 until  r.length){
   val j=(r.apply(i)+"").split('\t')(1);
   if(!"IPError".equals(isIP(j))){
     writer.write(get(j)+ "\n")
     Thread.sleep(1000)
   }
  }
  writer.close()
}

def isIP(str : String): String ={
  val pattern = "(\\d+)\\.(\\d+)\\.(\\d+)\\.(\\d+)".r

  matchTest(pattern findFirstIn str)
}

def matchTest(x: Any): String = x match {
  case Some(a) => a.toString
  case _ => "IPError"
}

  //请求淘宝接口
  def get(ip: String,
        connectTimeout: Int = 50000,
        readTimeout: Int = 50000,
        requestMethod: String = "GET") :String =
  {
    val urls="http://ip.taobao.com/service/getIpInfo.php?ip="+ip
    import java.net.{HttpURLConnection, URL}
    val connection = (new
URL(urls)).openConnection.asInstanceOf[HttpURLConnection]
    connection.setConnectTimeout(connectTimeout)
    connection.setReadTimeout(readTimeout)
    connection.setRequestMethod(requestMethod)
    val inputStream = connection.getInputStream
    val content = Source.fromInputStream(inputStream).mkString
    if (inputStream != null) inputStream.close
   println(content)
   content

  }
}
```

综合题执行结果如图 8-43 所示。

```
{"code":0,"data":{"country":"中国","country_id":"CN","area":"华北","area_id":"100000","region":"天津市","region_id":"120000",
"city":"天津市","city_id":"120100","county":"","county_id":"-1","isp":"电信","isp_id":"100017","ip":"123.150.156.4"}}
{"code":0,"data":{"country":"中国","country_id":"CN","area":"华东","area_id":"300000","region":"福建省","region_id":"350000",
"city":"福州市","city_id":"350100","county":"","county_id":"-1","isp":"电信","isp_id":"100017","ip":"110.84.190.37"}}
```

图 8-43　综合题执行结果

8.14　小结

从第 3~7 章读者基本熟悉了 Apache 版本的 Hadoop 集群，从本章开始我们把 Hadoop 开发平台改为 CDH 版本进行讲解。

本章主要包括 3 个部分。第一部分讲解 Scala 的简介、安装、Scala 开发环境的搭建；第二部分讲解 Scala 的语法，包括：基础语法、函数、控制语句、函数式编程、模式匹配、类和对象、Scala 异常处理、Trait（特征）、Scala 的 I/O；第三部分通过 5 道 Scala 的练习题目将 Scala 所有知识点进行整合，让读者在练习中复习和巩固相关的知识点。

第 9 章 Flume实战

9.1 Flume 概述

Flume 是由 Cloudera 软件公司开发出来的可分布式日志收集系统，它于 2009 年被捐赠了 Apache 软件基金会，为 Hadoop 相关组件之一。尤其近几年随着 Flume 的不断被完善以及升级版本的逐一推出，特别是 flume-ng 版本的开发，同时 Flume 内部的各种组件不断丰富，用户在开发的过程中使用的便利性得到很大的改善，现已成为 Apache 顶级项目之一。

1. 什么是 Flume

Apache Flume 是一个可以收集（例如日志、事件等）数据资源，并将这些数量庞大的数据从各项数据资源中集中起来存储的工具/服务，或者数据集中机制。Flume 具有高可用、分布式、可配置的工具，其设计的原理也是基于将数据流（如日志数据）从各种网站服务器上汇集起来，存储到 HDFS、HBase 等集中存储器中。Flume 结构如图 9-1 所示。

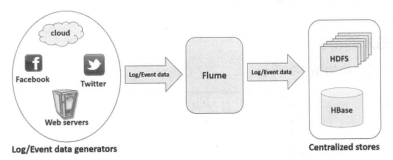

图 9-1 Flume 结构

2. 应用场景

比如我们在做一个电子商务网站，我们想根据消费用户中访问点特定的节点区域来分析消费者的行为或者购买意图。这样就可以更加快速地将他想要的商品推送到界面上，为了实现这一点，我们需要收集到他访问的页面以及单击的产品数据等日志数据信息，并移交给 Hadoop 平台上去分析，而 Flume 可以帮我们做到这一点。现在流行的内容推送，比如广告定点投放以及新闻私人定制也是基于此，不过不一定是使用 Flume，毕竟优秀的产品很多，比如 Facebook 的 Scribe、Apache 新出的另一个明星项目 Chukwa，还有淘宝 Time Tunnel。

3. Flume 的优势

（1）Flume 可以将应用产生的数据存储到任何集中存储器中，比如 HDFS、HBase。

（2）当收集数据的速度超过将写入数据速度的时候，也就是当收集信息遇到峰值时，收集的信息非常大，甚至超过了系统的写入数据能力，这时候 Flume 会在数据生产者和数据收容器间做出调整，保证其能够在两者之间提供一种平稳的数据。

（3）提供上下文路由特征。

（4）Flume 的管道是基于事务，保证了数据在传送和接收时的一致性。

（5）Flume 是可靠的、容错性高的、可升级的、易管理的，并且是可定制的。

4. Flume 具有的特征

（1）Flume 可以高效地将多个网站服务器中收集的日志信息存入 HDFS/HBase 中。

（2）使用 Flume，我们可以以将从多个服务器中获取的数据迅速地移交给 Hadoop 处理。

（3）除了日志信息，Flume 同时也可以用来接入收集规模宏大的社交网络节点事件数据（比如 Facebook、Twitter）、电商网站（如亚马逊、Flipkart）等。

（4）支持各种接入资源数据的类型以及接出数据类型。

（5）支持多路径流量、多管道接入流量、多管道接出流量、上下文路由等。

（6）可以被水平扩展。

9.2 Flume 的结构

1. Flume 的外部结构

Flume 的外部结构如图 9-2 所示。

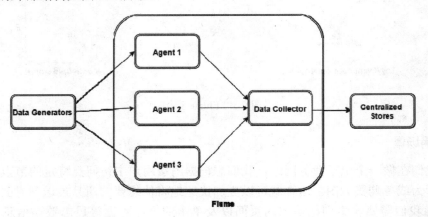

图 9-2　Flume 的外部结构

如图 9-2 所示，数据发生器（如 Facebook、Twitter）产生的数据被单个地运行在数据发生器所在服务器上的 Agent 所收集，之后数据收集器从各个 Agent 上汇集数据，并将采集到的数据存入到 HDFS 或者 HBase 中。

2. Flume 事件

事件作为 Flume 内部数据传输的最基本单元。它是由一个转载数据的字节数组（该数据组是从数据源接入点传入，并传输给传输器，也就是 HDFS/HBase）和一个可选头部构成。

典型的 Flume 事件结构如图 9-3 所示。

图 9-3 Flume 事件结构

在私人定制插件时，比如定制 flume-hbase-sink 插件，获取的就是 event，然后对其解析，并依据情况做过滤等，然后再传输给 HBase 或者 HDFS。

3. Flume Agent

了解了 Flume 的外部结构之后，知道了 Flume 内部有一个或者多个 Agent，对于每一个 Agent 来说，它就是一个独立的守护进程（JVM），它从客户端接收数据，或者从其他的 Agent 接收，然后迅速地将获取的数据传给下一个目的节点 Sink，或者 Agent。如图 9-4 所示是 Flume 的基本模型。

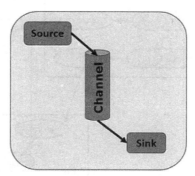

图 9-4 Flume 的基本模型

Agent 主要由 Source、Channel、Sink 三个组件组成。

（1）Source

Source 从数据发生器接收数据，并将接收的数据以 Flume 的 event 格式传递给一个或者多个通道 channal。Flume 提供多种数据接收的方式，比如 Avro、Thrift、twitter1% 等。

（2）Channel

Channal 是一种短暂的存储容器，它将从 Source 处接收到的 event 格式的数据缓存起来,直到它们被 Sinks 消费掉,它在 Source 和 Sink 间起着一个桥梁的作用,Channal 是一个完整的事务,这一点保证了数据在收发时候的一致性,并且它可以和任意数量的 Source 和 Sink 链接。支持的类型有 JDBC Channel、File System Channel、Memort Channel 等。

(3) Sink

Sink 将数据存储到集中存储器,比如 HBase 和 HDFS,它从 Channals 消费数据(events)并将其传递给目标地。目标地可能是另一个 Sink,也可能 HDFS、HBase。

Source、Channel、Sink 的组合形式例子如图 9-5 所示和图 9-6 所示。

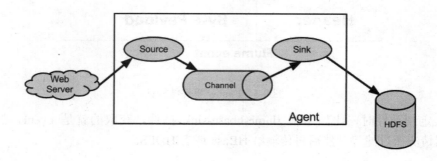

图 9-5 Source、Channel、Sink 的组合形式一

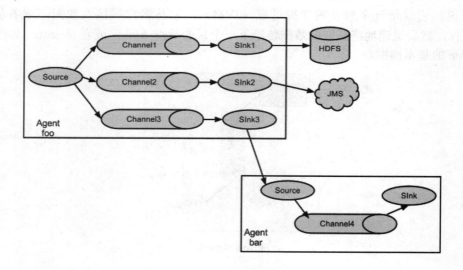

图 9-6 Source、Channel、Sink 的组合形式二

以上介绍的是 Flume 的主要组件,下面介绍两个次要组件。

(1) 拦截器

用于 Source 和 Channel 之间,用来更改或者检查 Flume 的 events 数据。

(2) 管道选择器

在多管道时被用来选择使用那一条管道来传递数据(events)。管道选择器又分为如下两种:

- 默认管道选择器：每一个管道传递的都是相同的 events。
- 多路复用通道选择器：依据每一个 event 的头部 header 的地址选择管道。

9.3 Flume 安装

本小节是 Flume 的 Apache 版本安装，在 CDH 版本只需要选择 Flume 组件安装。

1. Flume 下载

Flume 官网下载地址：https://flume.apache.org/download.html。

2. 安装

```
[root@hadoop11 ~]# cd /usr/app/
[root@hadoop11 app]# tar -zxvf apache-flume-1.6.0-bin.tar.gz
[root@hadoop11 app]# mv apache-flume-1.6.0-bin flume1.6
```

3. 配置

（1）配置/etc/profile，Flume 环境变量设置如图 9-7 所示。

```
export JAVA_HOME=/usr/app/jdk1.7
export HADOOP_HOME=/usr/app/hadoop-2.6.0
export HBASE_HOME=/usr/app/hbase
export HIVE_HOME=/usr/app/hive
export FLUME_HOME=/usr/app/flume1.6
export PATH=$PATH:$JAVA_HOME/bin:$HADOOP_HOME/bin:$HBASE_HOME/bin:$HIVE_HOME/bin:$FLUME_HOME/bin
```

图 9-7　Flume 环境变量设置

（2）配置/usr/app/flume1.6/conf/flume-env.sh。

```
[root@hadoop11 conf]# cp -r flume-env.sh.template flume-env.sh
[root@hadoop11 conf]# chmod 777 flume-env.sh
[root@hadoop11 conf]# vi flume-env.sh
```

4. flume-env.sh 添加 Java 路径

flume-env.sh 添加 Java 路径如图 9-8 所示。

```
# Enviroment variables can be set here.
export JAVA_HOME=/usr/app/jdk1.7
```

图 9-8　flume-env.sh 添加 Java 路径

Flume 组件 conf 目录下的配置文件如图 9-9 所示。

```
[root@hadoop11 conf]# ll
总用量 20
-rw-r--r--. 1 501   games 1661 5月   9 2015 flume-conf.properties.template
-rw-r--r--. 1 501   games 1110 5月   9 2015 flume-env.ps1.template
-rwxrwxrwx. 1 root  root  1204 4月  16 06:20 flume-env.sh
-rw-r--r--. 1 501   games 1214 5月   9 2015 flume-env.sh.template
-rw-r--r--. 1 501   games 3107 5月   9 2015 log4j.properties
[root@hadoop11 conf]#
```

图 9-9　Flume 组件 conf 目录下的配置文件

5. Flume 版本测试

Flume 版本测试如图 9-10 所示。

```
[root@hadoop11 app]# flume-ng version
Flume 1.6.0
Source code repository: https://git-wip-us.apache.org/repos/asf/flume.git
Revision: 2561a23240a71ba20bf288c7c2cda88f443c2080
Compiled by hshreedharan on Mon May 11 11:15:44 PDT 2015
From source with checksum b29e416802ce9ece3269d34233baf43f
[root@hadoop11 app]#
```

图 9-10　Flume 版本测试

9.4　Flume 实战

1. CDH 的配置方法

Flume 的 Agent 配置如图 9-11 所示，其中代理名称中写上配置名称，配置文件中填写具体配置内容。

图 9-11　Flume 的 Agent 配置

2. 本地到 Kafka 配置

本地到 Kafka 的配置文件如下：

```
#agent1 name
agent1.sources=source1
agent1.sinks=sink1
agent1.channels=channel1


#Spooling Directory
#set source1
agent1.sources.source1.type=spooldir
agent1.sources.source1.spoolDir=/home/yuhui/flumePath/dir/logdfs
agent1.sources.source1.channels=channel1
agent1.sources.source1.fileHeader = false
agent1.sources.source1.interceptors = i1
agent1.sources.source1.interceptors.i1.type = timestamp

#set sink1
agent1.sinks.sink1.type = org.apache.flume.sink.kafka.KafkaSink
agent1.sinks.sink1.topic = topicTest
agent1.sinks.sink1.brokerList = tagtic-slave01:9092,tagtic-slave02:9092,tagtic-slave03:9092
agent1.sinks.sink1.requiredAcks = 1
agent1.sinks.sink1.batchSize = 100
agent1.sinks.sink1.channel = channel1

#set channel1
agent1.channels.channel1.type=file
agent1.channels.channel1.checkpointDir=/home/yuhui/flumePath/dir/logdfstmp/point
agent1.channels.channel1.dataDirs=/home/yuhui/flumePath/dir/logdfstmp
```

3. 本地到 HDFS 配置

本地到 HDFS 的配置如下：

```
#agent1 name
agent1.sources=source1
agent1.sinks=sink1
agent1.channels=channel1

#Spooling Directory
#set source1
agent1.sources.source1.type=spooldir
agent1.sources.source1.spoolDir=/usr/app/flumelog/dir/logdfs
agent1.sources.source1.channels=channel1
agent1.sources.source1.fileHeader = false
agent1.sources.source1.interceptors = i1
```

```
agent1.sources.source1.interceptors.i1.type = timestamp

#set sink1
agent1.sinks.sink1.type=hdfs
agent1.sinks.sink1.hdfs.path=/user/yuhui/flume
agent1.sinks.sink1.hdfs.fileType=DataStream
agent1.sinks.sink1.hdfs.writeFormat=TEXT
agent1.sinks.sink1.hdfs.rollInterval=1
agent1.sinks.sink1.channel=channel1
agent1.sinks.sink1.hdfs.filePrefix=%Y-%m-%d

#set channel1
agent1.channels.channel1.type=file
agent1.channels.channel1.checkpointDir=/usr/app/flumelog/dir/logdfstmp/point
agent1.channels.channel1.dataDirs=/usr/app/flumelog/dir/logdfstmp
```

9.5 小结

Flume 是 Cloudera 提供的一个高可用的、高可靠的、分布式的海量日志采集、聚合和传输的系统。本章主要是以配置为主进行讲解：首先讲解 Flume 的背景、概述、结构，其次讲解 Apache 平台搭建 Flume，最后描述 CDH 和 Apache 两个版本 Hadoop 环境下的 Flume 配置。

第 10 章 Kafka 实战

10.1 Kafka 概述

10.1.1 简介

按照官方的说法：Kafka is a distributed,partitioned,replicated commit logservice。它提供了类似于 JMS 的特性，但是在设计实现上完全不同，此外它并不是 JMS 规范的实现。Kafka 在消息保存时根据 Topic 进行归类，消息发送者成为 Producer，消息接受者成为 Consumer，此外 Kafka 集群由多个 Kafka 实例组成，每个实例（Server）成为 Broker（每台机器）。无论是 Kafka 集群，还是 Producer 和 Consumer 都依赖于 ZooKeeper 来保证系统可用性，而集群保存一些 meta 信息。Kafka 集群消费者和生产者如图 10-1 所示。

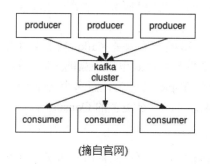

（摘自官网）

图 10-1　Kafka 集群消费者和生产者

1. Topics 和 logs

一个 Topic 可以认为是一类消息，每个 Topic 将被分成多个 partition（区），每个 partition 在存储层面是 append log 文件（追加）。任何发布到此 partition 的消息都会被直接追加到 log 文件的尾部，每条消息在文件中的位置称为 offset（偏移量），offset 为一个 long 型数字，它唯一地标记一条消息。Kafka 并没有提供其他额外的索引机制来存储 offset，因为在 Kafka 中几乎不允许对消息进行"随机读写"。Topics 和 logs 关联如图 10-2 所示。

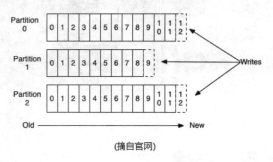

图 10-2　Topics 和 logs 关联

Kafka 和 JMS（Java Message Service）实现（activeMQ）不同的是：即使消息被消费，消息仍然不会被立即删除。日志文件将会根据 Broker 中的配置要求，留一定的时间之后删除，比如 log 文件保留 2 天，那么两天后文件会被清除，无论其中的消息是否被消费。Kafka 通过这种简单的手段来释放磁盘空间，以及减少消息消费之后对文件内容改动的磁盘 IO 开支。

对于 Consumer 而言，它需要保存消费消息的 offset，对于 offset 的保存和使用，由 Consumer 来控制。当 Consumer 正常消费消息时，offset 将会"线性"地向前驱动，即消息将依次被消费。事实上 Consumer 可以使用任意顺序消费消息，它只需要将 offset 重置为任意值。

Kafka 集群几乎不需要维护任何 Consumer 和 Producer 状态信息，这些信息由 ZooKeeper 保存。因此 Producer 和 Consumer 的客户端实现非常轻量级，它们可以随意离开，而不会对集群造成额外的影响。

Partitions 的设计目的有多个，最根本原因是 Kafka 基于文件存储。通过分区，可以将日志内容分散到多个 Server 上来避免文件尺寸达到单机磁盘的上限，每个 Partition 都会被当前 Server（Kafka 实例）保存。可以将一个 Topic 切分为任意多个 Partitions，来提高消息保存或消费的效率。此外，越多的 Partitions 意味着可以容纳更多的 Consumer，有效地提升并发消费的能力。

2. Distribution

一个 Topic 的多个 Partitions，被分布在 Kafka 集群中的多个 Server 上。每个 Server（Kafka 实例）负责 Partitions 中消息的读写操作。此外 Kafka 还可以配置 Partitions 需要备份的个数（replicas），每个 Partition 将会被备份到多台机器上，以提高可用性。

如果使用基于 replicated 方案，那么就意味着需要对多个备份进行调度。每个 Partition 都有一个 Server 为 leader。leader 负责所有的读写操作，如果 leader 失效，那么将会有其他 follower 来接管（成为新的 leader）。follower 只是单调地和 leader 跟进，同步消息即可。由此可见，作为 leader 的 Server 承载了全部的请求压力，因此从集群的整体考虑，有多少个 Partitions 就意味着有多少个 leader，Kafka 会将 leader 均衡地分散在每个实例上，来确保整体的性能稳定。

（1）Producers

Producer 将消息发布到指定的 Topic 中，同时 Producer 也能决定将此消息归属于哪个 partition。比如基于 round-robin 方式或者通过其他的一些算法等。

（2）Consumers

本质上 Kafka 只支持 Topic。每个 Consumer 属于一个 Consumer group，反过来说，每个 group 中可以有多个 Consumer。发送到 Topic 的消息，只会被订阅此 Topic 的每个 group 中的一个 Consumer 消费。

如果所有的 Consumer 都具有相同的 group，这种情况和 queue 模式很像。消息将会在 Consumers 之间负载均衡。

如果所有的 Consumer 都具有不同的 group，那就是"发布-订阅"。消息将会广播给所有的消费者。

在 Kafka 中，一个 Partition 中的消息只会被 group 中的一个 Consumer 消费。每个 group 中 Consumer 消息消费互相独立。可以认为一个 group 是一个"订阅者"，一个 Topic 中的每个 Partitions，只会被一个"订阅者"中的一个 Consumer 消费，不过一个 Consumer 可以消费多个 partitions 中的消息。Kafka 只能保证一个 Partition 中的消息被某个 Consumer 消费时，消息是顺序的。事实上，从 Topic 角度来说，消息仍不是有序的。

Kafka 的设计原理决定，对于一个 topic，同一个 group 中不能有多于 Partitions 个数的 Consumer 同时消费，否则将意味着某些 Consumer 将无法得到消息。

（3）Guarantees

发送到 Partitions 中的消息将会按照它接收的顺序追加到日志中。

对于消费者而言，它们消费消息的顺序和日志中消息顺序一致。

如果 Topic 的 replicationfactor 为 N，那么允许 N-1 个 Kafka 实例失效。

10.1.2　使用场景

1. Messaging

对于一些常规的消息系统，Kafka 是个不错的选择。Partitions/replication 和容错，可以使 Kafka 具有良好的扩展性和性能优势。不过到目前为止，我们应该很清楚地认识到，Kafka 并没有提供 JMS 中的"事务性""消息传输担保（消息确认机制）""消息分组"等企业级特性。Kafka 只能使用作为"常规"的消息系统，在一定程度上，尚未确保消息的发送与接收绝对可靠（比如，消息重发、消息发送丢失等）。

2. Websit Activity Tracking

Kafka 可以作为"网站活性跟踪"的最佳工具，可以将网页/用户操作等信息发送到 Kafka 中，并实时监控，或者离线统计分析等。

3. Log Aggregation

Kafka 的特性决定了它非常适合作为日志收集中心。Application 可以将操作日志"批量""异步"的发送到 Kafka 集群中，而不是保存在本地或者 DB 中。Kafka 可以批量提交消息/压缩消息等，这对 Producer 端而言，几乎感觉不到性能的开支。此时 Consumer 端可以使用 Hadoop 等其他系统化的存储和分析系统。

10.2 Kafka 设计原理

Kafka 的设计初衷是希望作为一个统一的信息收集平台,能够实时地收集反馈信息,并要求能够支撑较大的数据量,且具备良好的容错能力。

1. 持久性

Kafka 使用文件存储消息,这就直接决定 Kafka 在性能上严重依赖文件系统的本身特性,且无论任何 OS 下,对文件系统本身的优化几乎没有可能。文件缓存/直接内存映射等是常用的手段。因为 Kafka 是对日志文件进行 append 操作,因此磁盘检索的开支是较小的。同时为了减少磁盘写入的次数,broker 会将消息暂时 buffer 起来,当消息的个数(或尺寸)达到一定阀值时,再 flush 到磁盘,这样减少了磁盘 IO 调用的次数。

2. 性能

需要考虑的影响性能点很多,除磁盘 IO 之外,还需要考虑网络 IO,这直接关系到 Kafka 的吞吐量问题。Kafka 并没有提供太多高超的技巧。对于 Producer 端,可以将消息 buffer 起来,当消息的条数达到一定阀值时,批量发送给 broker。对于 Consumer 端也是一样,批量 fetch 多条消息。不过消息量的大小可以通过配置文件来指定。对于 Kafka broker 端,似乎有个 sendfile 系统调用可以潜在地提升网络 IO 的性能:将文件的数据映射到系统内存中,socket 直接读取相应的内存区域即可,而无须进程再次复制和交换。其实对于 Producer/Consumer/Broker 三者而言,CPU 的开支应该都不大,因此启用消息压缩机制是一个良好的策略。压缩需要消耗少量的 CPU 资源,不过对于 Kafka 而言,网络 IO 更应该需要考虑。可以将任何在网络上传输的消息都经过压缩。Kafka 支持 gzip/snappy 等多种压缩方式。

3. 生产者

负载均衡:Producer 将会和 Topic 下所有 Partition leader 保持 socket 连接。消息由 Producer 直接通过 socket 发送到 Broker,中间不会经过任何"路由层"。事实上,消息被路由到哪个 Partition 上,由 Producer 客户端决定。比如可以采用"random""key-hash""轮询"等,如果一个 Topic 中有多个 partitions,那么在 Producer 端实现"消息均衡分发"是必要的。

其中,Partition leader 的位置(host:port)注册在 ZooKeeper 中,Producer 作为 ZooKeeper client,已经注册了 watch 用来监听 Partition leader 的变更事件。

异步发送:将多条消息暂且在客户端 buffer 起来,并将他们批量地发送到 Broker,小数据 IO 太多,会拖慢整体的网络延迟,批量延迟发送事实上提升了网络效率。不过这也有一定的隐患,比如说当 Producer 失效时,那些尚未发送的消息将会丢失。

4. 消费者

Consumer 端向 Broker 发送 fetch 请求,并告知其获取消息的 offset。此后 Consumer 将会获得一定条数的消息。Consumer 端也可以重置 offset 来重新消费消息。

在 JMS 实现中,Topic 模型基于 push 方式,即 Broker 将消息推送给 Consumer 端。不过在 Kafka 中，采用了 pull 方式，即 Consumer 在和 Broker 建立连接之后，主动去 pull（或者说 fetch）消息。这种模式有其优点：首先 Consumer 端可以根据自己的消费能力适时地去 fetch 消息并处理，且可以控制消息消费的进度（offset）。此外，消费者可以良好地控制消息消费的数量 batch fetch。

其他 JMS 实现，消息消费的位置是由 prodiver 保留，以便避免重复发送消息或者将没有消费成功的消息重发等，同时还要控制消息的状态。这就要求 JMS Broker 需要太多额外的工作。在 Kafka 中，Partition 中的消息只有一个 Consumer 在消费，且不存在消息状态的控制，也没有复杂的消息确认机制，可见 Kafka Broker 端是相当轻量级的。当消息被 Consumer 接收之后，Consumer 可以在本地保存最后消息的 offset，并间歇性地向 ZooKeeper 注册 offset。由此可见，Consumer 客户端也很轻量级。Kafka 消费者如图 10-3 所示。

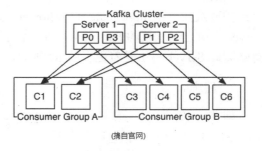

图 10-3 Kafka 消费者

5. 消息传送机制

对于 JMS 实现，消息传输担保非常直接：有且只有一次（exactly once）。在 Kafka 中稍有不同。

（1）at most once：最多一次，这个和 JMS 中"非持久化"消息类似。发送一次，无论成败，都不会重发。

（2）at least once：消息至少发送一次，如果消息未能接受成功，可能会重发，直到接收成功。

（3）exactly once：消息只会发送一次。

at most once：消费者 fetch 消息，然后保存 offset，然后处理消息。当 Client 保存 offset 之后，但是在消息处理过程中出现了异常，导致部分消息未能继续处理。那么此后"未处理"的消息将不能被 fetch 到，这就是 at most once。

at least once：消费者 fetch 消息，然后处理消息，然后保存 offset。如果消息处理成功之后，但是在保存 offset 阶段 ZooKeeper 异常导致保存操作未能执行成功，这就导致接下来再次 fetch 时可能获得上次已经处理过的消息，这就是 at least once，原因 offset 没有及时提交给 ZooKeeper，ZooKeeper 恢复正常还是之前 offset 状态。

exactly once：Kafka 中并没有严格地去实现（基于二阶段提交，事务），我们认为这种策略在 Kafka 中是没有必要的。

通常情况下 at least once 是我们的首选。相比 at most once 而言，重复接收数据总比丢失数据要好。

6. 复制备份

Kafka 将每个 Partition 数据复制到多个 Server 上，任何一个 Partition 有一个 leader 和多个 follower（可以没有）。备份的个数可以通过 Broker 配置文件来设定。leader 处理所有的

read-write 请求，follower 需要和 leader 保持同步。follower 和 Consumer 一样，消费消息并保存在本地日志中。leader 负责跟踪所有的 follower 状态，如果 follower "落后"太多或者失效，leader 将会把它从 replicas 同步列表中删除。当所有的 follower 都将一条消息保存成功，此消息才被认为是"committed"，那么此时 Consumer 才能消费它。即使只有一个 replicas 实例存活，仍然可以保证消息的正常发送和接收，只要 ZooKeeper 集群存活即可（不同于其他分布式存储，比如 HBase 需要"多数派"存活才行）。

当 leader 失效时，需在 followers 中选取出新的 leader，可能此时 follower 落后于 leader，因此需要选择一个 up-to-date 的 follower。选择 follower 时需要兼顾一个问题，就是新 leader Server 上所已经承载的 Partition leader 的个数，如果一个 Server 上有过多的 Partition leader，意味着此 Server 将承受着更多的 IO 压力。在选举新 leader 的，需要考虑到负载均衡。

7. 日志

如果一个 Topic 的名称为 my_topic，它有 2 个 Partitions，那么日志将会保存在 my_topic_0 和 my_topic_1 两个目录中。日志文件中保存了一个序列 log entries（日志条目），每个 log entry 格式为"4 个字节的数字 N 表示消息的长度 + N 个字节的消息内容"。每个日志都有一个 offset 来唯一地标记一条消息，offset 的值为 8 个字节的数字，表示此消息在 Partition 中所处的起始位置。每个 Partition 在物理存储层面，由多个 log file 组成（称为 segment）。segmentfile 的命名为"最小 offset.kafka"，例如"00000000000.kafka"，其中"最小 offset"表示此 segment 中起始消息的 offset。Kafka 日志结构如图 10-4 所示。

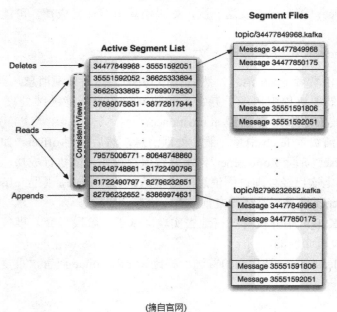

（摘自官网）

图 10-4　Kafka 日志结构

其中每个 Partition 中所持有的 segments 列表信息会存储在 ZooKeeper 中。

当 segment 文件尺寸达到一定阀值时（可以通过配置文件设定，默认 1GB），将会创建一个新的文件。当 buffer 中消息的条数达到阀值时，将会触发日志信息 flush 到日志文件中，同时如果"距离最近一次 flush 的时间差"达到阀值时，也会触发 flush 到日志文件。如果 Broker 失效，极有可能会丢失那些尚未 flush 到文件的消息。因为 Server 意外实现，仍然会导致 log 文件格式的破坏（文件尾部），那么就要求当 Server 启动时，需要检测最后一个 segment 的文件结构是否合法并进行必要的修复。

获取消息时，需要指定 offset 和最大 chunk 尺寸，offset 用来表示消息的起始位置，chunk size 用来表示最大获取消息的总长度（间接地表示消息的条数）。根据 offset，可以找到此消息所在 segment 文件，然后根据 segment 的最小 offset 取差值，得到它在 file 中的相对位置，直接读取输出即可。

日志文件的删除策略非常简单：启动一个后台线程定期扫描 log file 列表，把保存时间超过阀值的文件直接删除（根据文件的创建时间）。为了避免删除文件时仍然有 read 操作（Consumer 消费），采取 copy-on-write 方式。

8. 分配

Kafka 使用 ZooKeeper 来存储一些 meta 信息，并使用了 ZooKeeper watch 机制来发现 meta 信息的变更并作出相应的动作（比如 Consumer 失效、触发负载均衡等）。

（1）Broker node registry：当一个 Kafka Broker 启动后，首先会向 ZooKeeper 注册自己的节点信息（临时 znode），同时当 Broker 和 ZooKeeper 断开连接时，此 znode 也会被删除。

格式：/broker/ids/[0...N]-->host:port;

其中[0..N]表示 Broker id，每个 Broker 的配置文件中都需要指定一个数字类型的 id（全局不可重复），znode 的值为此 Broker 的 host:port 信息。

（2）Broker Topic Registry：当一个 Broker 启动时，会向 ZooKeeper 注册自己持有的 topic 和 partitions 信息，仍然是一个临时 znode。

格式：/broker/topics/[topic]/[0...N]

其中[0..N]表示 partition 索引号。

（3）Consumer and Consumer group：每个 Consumer 客户端被创建时，会向 ZooKeeper 注册自己的信息，此作用主要是为了"负载均衡"。

一个 group 中的多个 Consumer 可以交错地消费一个 Topic 的所有 Partitions。简而言之，保证此 Topic 的所有 Partitions 都能被此 group 所消费，且消费时为了性能考虑，让 Partition 相对均衡地分散到每个 Consumer 上。

（4）Consumer id Registry：每个 Consumer 都有一个唯一的 ID（host:uuid，可以通过配置文件指定，也可以由系统生成），此 ID 用来标记消费者信息。

格式：/consumers/[group_id]/ids/[consumer_id]

仍然是一个临时的 znode，此节点的值为{"topic_name":#streams...}，即表示此 Consumer 目前所消费的 topic + Partitions 列表。

（5）Consumer offset Tracking：用来跟踪每个 Consumer 目前所消费的 Partition 中最大的 offset。

格式：/consumers/[group_id]/offsets/[topic]/[broker_id-partition_id]-->offset_value

此 znode 为持久节点，可以看出 offset 跟 group_id 有关，以表明当 group 中一个消费者失效，其他 Consumer 可以继续消费。

（6）Partition Owner registry：用来标记 Partition 被哪个 Consumer 消费临时 znode。

格式：/consumers/[group_id]/owners/[topic]/[broker_id-partition_id]-->consumer_node_id

当 Consumer 启动时，所触发的操作：

（1）首先进行"Consumer id Registry"。

（2）然后在"Consumer id Registry"节点下注册一个 watch 用来监听当前 group 中其他 Consumer 的 leave 和 join。只要此 znode path 下节点列表变更，都会触发此 group 下 Consumer 的负载均衡，比如一个 Consumer 失效，那么其他 Consumer 接管 Partitions。

（3）在"Broker id registry"节点下，注册一个 watch 用来监听 Broker 的存活情况。如果 Broker 列表变更，将会触发所有的 groups 下的 Consumer 重新 balance。

Producer、Consumer、ZooKeeper 关系执行步骤如图 10-5 所示。

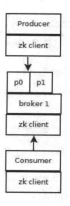

图 10-5 Producer、Consumer、ZooKeeper 关系执行步骤

（1）Producer 端使用 ZooKeeper 来"发现"Broker 列表，以及和 Topic 下每个 Partition leader 建立 socket 连接并发送消息。

（2）Broker 端使用 ZooKeeper 来注册 Broker 信息，以及监测 Partition leader 存活性。

（3）Consumer 端使用 ZooKeeper 来注册 Consumer 信息，其中包括 Consumer 消费的 Partition 列表等，同时也用来发现 Broker 列表，并和 Partition leader 建立 socket 连接，并获取消息。

10.3 Kafka 主要配置

1. Broker 主要配置

Broker 主要配置如图 10-6 所示。

```
1.  ##broker标识,id为正数,且全局不得重复.
2.  broker.id=1
3.  ##日志文件保存的目录
4.  log.dirs=~/kafka/logs
5.  ##broker需要使用zookeeper保存meata信息,因此broker为zk client;
6.  ##此处为zookeeper集群的connectString,后面可以跟上path,比如
7.  ##hostname:port/chroot/kafka
8.  ##不过需要注意,path的全路径需要有自己来创建(使用zookeeper脚本工具)
9.  zookeeper.connect=hostname1:port1,hostname2:port2
10. ##用来侦听链接的端口,prudcer或consumer将在此端口建立链接
11. port=6667
12. ##指定broker实例绑定的网络接口地址
13. host.name=
14. ##每个partition的备份个数,默认为1,建议根据实际条件选择
15. ##此致值大意味着消息各个server上同步时需要的延迟较高
16. num.partitions=2
17. ##日志文件中每个segment文件的尺寸,默认为1G
18. ##log.segment.bytes=1024*1024*1024
19. ##滚动生成新的segment文件的最大时间
20. ##log.roll.hours=24*7
21. ##segment文件保留的最长时间,超时将被删除
22. ##log.retention.hours=24*7
23. ##partiton中buffer中,消息的条数,达到阀值,将触发flush到磁盘.
24. log.flush.interval.messages=10000
25. ##消息buffer的时间,达到阀值,将触发flush到磁盘.
26. log.flush.interval.ms=3000
27. ##partition leader等待follower同步消息的最大时间,
28. ##如果超时,leader将follower移除同步列表
29. replica.lag.time.max.ms=10000
30. ##允许follower落后的最大消息条数,如果达到阀值,将follower移除同步列表
31. ##replica.lag.max.message=4000
32. ##消息的备份的个数,默认为1
33. num.replica.fetchers=1
```

图 10-6 Broker 主要配置

2. Consumer 主要配置

Consumer 主要配置如图 10-7 所示。

```
1.  ##当前消费者的group名称,需要指定
2.  group.id=
3.  ##consumer作为zookeeper client,需要通过zk保存一些meta信息,此处为zk connectString
4.  zookeeper.connect=hostname1:port,hostname2:port2
5.  ##当前consumer的标识,可以设定,也可以有系统生成
6.  conusmer.id=
7.  ##获取消息的最大尺寸,broker不会像consumer输出大于此值的消息chunk
8.  ##每次feth将得到多条消息,此值为总大小
9.  fetch.messages.max.bytes=1024*1024
10. ##当consumer消费一定量的消息之后,将会自动向zookeeper提交offset信息
11. ##注意offset信息并不是每消费一次消息,就像zk提交一次,而是现在本地保存,并定期提交
12. auto.commit.enable=true
13. ##自动提交的时间间隔,默认为1分钟.
14. auto.commit.interval.ms=60*1000
```

图 10-7 Consumer 主要配置

3. Producer 主要配置

Producer 主要配置如图 10-8 所示。

```
1.   ##对于开发者而言,需要通过broker.list指定当前producer需要关注的broker列表
2.   ##producer通过和每个broker链接,并获取partitions,
3.   ##如果某个broker链接失败,将导致此上的partitons无法继续发布消息
4.   ##格式:host1:port,host2:port2,其中host:port需要参考broker配置文件.
5.   ##对于producer而言没有使用zookeeper自动发现broker列表,非常奇怪。(0.8V和0.7有区别)
6.   metadata.broker.list=
7.   ##producer接收消息ack的时机.默认为0.
8.   ##0:  producer不会等待broker发送ack
9.   ##1:  当leader接收到消息之后发送ack
10.  ##2:  当所有的follower都同步消息成功后发送ack.
11.  request.required.acks=0
12.  ##producer消息发送的模式,同步或异步.
13.  ##异步意味着消息将会在本地buffer,并适时批量发送
14.  ##默认为sync,建议async
15.  producer.type=sync
16.  ##消息序列化类,将消息实体转换成byte[]
17.  serializer.class=kafka.serializer.DefaultEncoder
18.  key.serializer.class=${serializer.class}
19.  ##partitions路由类,消息在发送时将根据此实例的方法获得partition索引号.
20.  partitioner.class=kafka.producer.DefaultPartitioner
21.
22.  ##消息压缩算法,none,gzip,snappy
23.  compression.codec=none
24.  ##消息在producer端buffer的条数.仅在producer.type=async下有效
25.  ##batch.num.messages=200
```

图 10-8 Producer 主要配置

10.4 Kafka 客户端操作

Kafka 查看 TopicName 的列表命令,如图 10-9 所示。

```
kafka-topics --list --zookeeper localhost:2181
```

```
[root@tagtic-master ~]#
[root@tagtic-master ~]# kafka-topics --list --zookeeper localhost:2181
```

图 10-9 Kafka 查看 topicName 的列表命令

Kafka 创建 Topic 命令,如图 10-10 所示。

```
kafka-topics --create --topic TestKafka --replication-factor 1 --partitions 10 --zookeeper localhost:2181
```

```
[root@tagtic-master ~]#
[root@tagtic-master ~]# kafka-topics --create --topic TestKafka --replication-factor 1 --partitions 10 --zookeeper localhost:2181
```

图 10-10 Kafka 创建 topic 命令

Kafka 生产者命令,如图 10-11 所示。

```
kafka-console-producer --broker-list tagtic-slave01:9092 --sync --topic TestKafka
```

```
[root@tagtic-master ~]#
[root@tagtic-master ~]# kafka-console-producer --broker-list tagtic-slave01:9092 --sync
--topic TestKafka
```

图 10-11　Kafka 生产者命令

Kafka 消费者命令，如图 10-12 所示。

```
kafka-console-consumer --zookeeper tagtic-master:2181 --topic
TestKafka--from-beginning
```

```
[root@tagtic-slave01 yuhui]#
[root@tagtic-slave01 yuhui]# kafka-console-consumer --zookeeper tagtic-master:2181
--topic TestKafka --from-beginning
```

图 10-12　Kafka 消费者命令

Kafka 查看指定主题命令，如图 10-13、图 10-14 所示。

```
kafka-topics --describe --zookeeper tagtic-master:2181 --topic
TestKafka
```

```
[root@tagtic-slave01 yuhui]#
[root@tagtic-slave01 yuhui]# kafka-topics --describe --zookeeper tagtic-master:2181
--topic TestKafka
```

图 10-13　Kafka 查看指定主题命令

```
17/05/19 08:44:30 INFO zkclient.ZkClient: zookeeper state changed (SyncConnected)
Topic:TestKafka  PartitionCount:10   ReplicationFactor:1   Configs:
        Topic: TestKafka    Partition: 0    Leader: 138    Replicas: 138    Isr: 138
        Topic: TestKafka    Partition: 1    Leader: 139    Replicas: 139    Isr: 139
        Topic: TestKafka    Partition: 2    Leader: 140    Replicas: 140    Isr: 140
        Topic: TestKafka    Partition: 3    Leader: 137    Replicas: 137    Isr: 137
        Topic: TestKafka    Partition: 4    Leader: 138    Replicas: 138    Isr: 138
        Topic: TestKafka    Partition: 5    Leader: 139    Replicas: 139    Isr: 139
        Topic: TestKafka    Partition: 6    Leader: 140    Replicas: 140    Isr: 140
        Topic: TestKafka    Partition: 7    Leader: 137    Replicas: 137    Isr: 137
        Topic: TestKafka    Partition: 8    Leader: 138    Replicas: 138    Isr: 138
        Topic: TestKafka    Partition: 9    Leader: 139    Replicas: 139    Isr: 139
17/05/19 08:44:30 INFO zkclient.ZkEventThread: Terminate ZkClient event thread.
17/05/19 08:44:30 INFO zookeeper.ZooKeeper: Session: 0x15be5987bf78b34 closed
17/05/19 08:44:30 INFO zookeeper.ClientCnxn: EventThread shut down
```

图 10-14　Kafka 查看指定主题

Kafka 删除 Topic 命令，如图 10-15 所示。

```
kafka-topics --zookeeper tagtic-master:2181 --topic TestKafka -
delete
```

```
[root@tagtic-slave01 yuhui]#
[root@tagtic-slave01 yuhui]# kafka-topics --zookeeper tagtic-master:2181 --topic
TestKafka --delete
```

图 10-15　Kafka 删除 topic 命令

10.5　Java 操作 Kafka

10.5.1　生产者

通过 Java 的 IO 读写本地文件 E:\\sdkJson.log。先将数据一条一条读取出来，接着使用 Kafka 的生产者 API 将数据一条一条发送到 TopicName 为 TestKafka 里面。

```java
package com.ou.cn;

import java.io.BufferedReader;
import java.io.FileReader;
import java.util.Properties;
import java.util.concurrent.TimeUnit;
import kafka.javaapi.producer.Producer;
import kafka.producer.KeyedMessage;
import kafka.producer.ProducerConfig;
import kafka.serializer.StringEncoder;


/***
 *
 * @author yuhui
 *
 * @data 2017年05月23日上午10:51:05
 */
public class kafkaProducer extends Thread{

    private String topic;

    public kafkaProducer(String topic){
        super();
        this.topic = topic;
    }

    @Override
    public void run() {
        Producer<Integer, String> producer = createProducer();
        int i=0;
        while(true){

            try {
                String path = "E:\\sdkJson.log";

                StringBuffer sb = null ;

                String line = null;
```

```java
            BufferedReader br = new BufferedReader(new FileReader(path)) ;

            while((line=br.readLine())!=null) {
                //往 Kafka 中生产数据
//                System.out.println(line);

                producer.send(new KeyedMessage<Integer, String>(topic, line));
            }

            System.out.println("文件读取完毕");
            Thread.sleep(1000L*60);
        } catch (Exception e1) {
            // TODO Auto-generated catch block
            e1.printStackTrace();
        }
        try {
            TimeUnit.SECONDS.sleep(1);
        } catch (InterruptedException e) {
            e.printStackTrace();
        }
      }
    }

    private Producer createProducer() {
        Properties properties = new Properties();
        properties.put("zookeeper.connect", "tagtic-slave01:2181,tagtic-slave02:2181,tagtic-slave03:2181");//声明 zk
        properties.put("serializer.class", StringEncoder.class.getName());
        properties.put("metadata.broker.list", "tagtic-slave01:9092,tagtic-slave02:9092,tagtic-slave03:9092");// 声明 Kafka broker
        properties.put("group.id", "group1");//声明用户组
        return new Producer<Integer, String>(new ProducerConfig(properties));
    }


    public static void main(String[] args) {
        new kafkaProducer("TestKafka").start();// 使用 Kafka 集群中创建好的主题 jdktest
    }

}
```

10.5.2 消费者

使用 Kafka 的消费者 API 将 TopicName 为 TestKafka 中的数据一条一条消费出来。

```java
package com.ou.cn;

import java.util.HashMap;
import java.util.List;
import java.util.Map;
import java.util.Properties;
import kafka.consumer.Consumer;
import kafka.consumer.ConsumerConfig;
import kafka.consumer.ConsumerIterator;
import kafka.consumer.KafkaStream;
import kafka.javaapi.consumer.ConsumerConnector;


/***
 *
 * @author yuhui
 *
 * @data 2017年06月23日上午10:50:55
 */
public class kafkaConsumer extends Thread{

    private String topic;

    public kafkaConsumer(String topic){
        super();
        this.topic = topic;
    }


    @Override
    public void run() {
        ConsumerConnector consumer = createConsumer();
        Map<String, Integer> topicCountMap = new HashMap<String, Integer>();
        topicCountMap.put(topic, 1); // 一次从主题中获取一个数据
        Map<String, List<KafkaStream<byte[], byte[]>>> messageStreams = consumer.createMessageStreams(topicCountMap);
        KafkaStream<byte[], byte[]> stream = messageStreams.get(topic).get(0);// 获取每次接收到的这个数据
        ConsumerIterator<byte[], byte[]> iterator = stream.iterator();
        int i=0;
        while(iterator.hasNext()){
            i++;
            String message = new String(iterator.next().message());
            System.out.println("消费出来的数据第   "+i +" 条是    : "+ message);
```

```java
        }
    }

    private ConsumerConnector createConsumer() {
        Properties properties = new Properties();
        properties.put("zookeeper.connect", "tagtic-slave01:2181,tagtic-slave02:2181,tagtic-slave03:2181");//声明zk
        properties.put("group.id", "group1");// 生产和消费必须要使用相同的组名称，如果生产者和消费者都不在同一组，则取不到数据
        return Consumer.createJavaConsumerConnector(new ConsumerConfig(properties));
    }

    public static void main(String[] args) {
        new kafkaConsumer("TestKafka").start();// 使用Kafka集群中创建好的主题 test
    }
}
```

10.6 Flume 连接 Kafka

本节给出一个 Flume 连接 Kafka 的实例测试，通过 Flume 组件将监控路径下面的文件存储到 Kafka 中，接着通过 Kafka 的客户端消费展示。

1. Kafka 的 topic 查看

```
Kafka-topics --list --zookeeper localhost:2181
```

2. Kafka 的 topic 建立

```
kafka-topics --create --topic topicTest --replication-factor 1 --partitions 10 --zookeeper localhost:2181
```

3. Flume 的监控和读写目录

```
[root@tagtic-slave03 yuhui]# mkdir /home/yuhui/flumePath
[root@tagtic-slave03 yuhui]# mkdir /home/yuhui/flumePath/dir
[root@tagtic-slave03 yuhui]# mkdir /home/yuhui/flumePath/dir/logdfs
[root@tagtic-slave03 yuhui]# mkdir /home/yuhui/flumePath/dir/logdfstmp
[root@tagtic-slave03 yuhui]# mkdir /home/yuhui/flumePath/dir /point
[root@tagtic-slave03 yuhui]# chmod 777 -R  /home/yuhui/flumePath/dir
```

4. Flume 的监控和读写目录解说

```
//日志监控路径，只要日志放入则被 Flume 监控
[root@tagtic-slave03 yuhui]# mkdir flumePath
[root@tagtic-slave03 flumePath]# pwd
/home/yuhui/flumePath
[root@tagtic-slave03 flumePath]# mkdir dir
[root@tagtic-slave03 flumePath]# mkdir dir/logdfs

//日志读取完毕存储路径，日志在这里则一直存储在 Channel 中（最多只有两个 log-number
//日志，且默认达到1.6G 之后删除前面一个 log，建立新的 log）
[root@tagtic-slave03 flumePath]# mkdir dir/logdfstmp

//日志监控路径中文件的路径存放点
[root@tagtic-slave03 flumePath]# mkdir dir/logdfstmp/point

//修改权限
[root@tagtic-slave03 flumePath]# chmod 777 -R  dir
```

5. Flume 的配置文件

```
#agent1 name
agent1.sources=source1
agent1.sinks=sink1
agent1.channels=channel1


#Spooling Directory
#set source1
agent1.sources.source1.type=spooldir
agent1.sources.source1.spoolDir=/home/yuhui/flumePath/dir/logdfs
agent1.sources.source1.channels=channel1
agent1.sources.source1.fileHeader = false
agent1.sources.source1.interceptors = i1
agent1.sources.source1.interceptors.i1.type = timestamp

#set sink1
agent1.sinks.sink1.type = org.apache.flume.sink.kafka.KafkaSink
agent1.sinks.sink1.topic = topicTest
agent1.sinks.sink1.brokerList = tagtic-slave01:9092,tagtic-slave02:9092,tagtic-slave03:9092
agent1.sinks.sink1.requiredAcks = 1
agent1.sinks.sink1.batchSize = 100
agent1.sinks.sink1.channel = channel1

#set channel1
agent1.channels.channel1.type=file
agent1.channels.channel1.checkpointDir=/home/yuhui/flumePath/dir/logdfstmp/point
agent1.channels.channel1.dataDirs=/home/yuhui/flumePath/dir/logdfstmp
```

6. 日志文件

将 sdkJson.log 文件移动到/home/yuhui/flumePath/dir/logdfs 路径下，让 Flume 读取 sdkJson.log 文件，将文件内容写入到 TopicName 为 topicTest 中。sdkJson.log 日志展示如图 10-16 所示。

```
[root@tagtic-slave03 yuhui]# mv  sdkJson.log
/home/yuhui/flumePath/dir/logdfs
```

图 10-16　sdkJson.log 日志展示

7. Kafka 的消费命令，消费 TopicName 为 topicTest 中的数据

Kafka 消费结果如图 10-17 所示。

```
kafka-console-consumer --zookeeper localhost:2181 --topic topicTest --from-beginning
```

图 10-17　Kafka 消费结果

10.7 小结

Kafka 是一种高吞吐量的分布式发布订阅消息系统。本章从 Kafka 的简介和使用场景开始，接着描述了设计原理、主要配置，之后实战方面通过 Kafka 的 Client 客户端和 Eclipse+Java 对 Kafka 进行操作，最后将 Flume 和 Kafka 进行整合，完成一个数据源测试。

第 11 章 Spark 实战

11.1 Spark 概述

Apache Spark 是专为大规模数据处理而设计的快速通用的计算引擎。

Spark 是 UC Berkeley AMP lab（加州大学伯克利分校的 AMP 实验室）所开源的类 Hadoop MapReduce 的通用并行框架，Spark 拥有 Hadoop MapReduce 所具有的优点；但不同于 MapReduce 的是 Job 中间输出结果可以保存在内存中，从而不再需要读写 HDFS，因此 Spark 能更好地适用于数据挖掘与机器学习等需要迭代的 MapReduce 的算法。

Spark 是一种与 Hadoop 相似的开源集群计算环境，但是两者之间还存在一些不同之处，这些有用的不同之处使 Spark 在某些工作负载方面表现得更加优越，换句话说，Spark 启用了内存分布数据集，除了能够提供交互式查询外，它还可以优化迭代工作负载。

Spark 是在 Scala 语言中实现的，它将 Scala 用作其应用程序框架。与 Hadoop 不同，Spark 和 Scala 能够紧密集成，其中的 Scala 可以像操作本地集合对象一样轻松地操作分布式数据集。

尽管创建 Spark 是为了支持分布式数据集上的迭代作业，但是实际上它是对 Hadoop 的补充，可以在 Hadoop 文件系统中并行运行。通过名为 Mesos 的第三方集群框架可以支持此行为。Spark 由加州大学伯克利分校 AMP 实验室（Algorithms, Machines and People Lab）开发，可用来构建大型的、低延迟的数据分析应用程序。

11.2 Spark 基本概念

1. Spark 特性
- 高可伸缩性
- 高容错
- 内存计算

2. Spark 的生态体系

Spark 属于 BDAS（BDAS，伯利克分析栈）生态体系。

- MapReduce 属于 Hadoop 生态体系之一，Spark 则属于 BDAS 生态体系之一。

- Hadoop 包含了 MapReduce、HDFS、HBase、Hive、ZooKeeper、Pig、Sqoop 等。
- BDAS 包含了 Spark、Shark（相当于 Hive）、BlinkDB、Spark Streaming（消息实时处理框架，类似 Storm）等。

BDAS 生态体系图如图 11-1 所示。

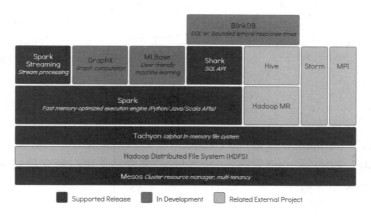

图 11-1　BDAS 生态体系

3. Spark 与 MapReduce

相对 MapReduce，Spark 具有如下优势：

- MapReduce 通常将中间结果放到 HDFS 上，Spark 是基于内存并行大数据框架，中间结果存放到内存，对于迭代数据 Spark 效率高。
- MapReduce 总是消耗大量时间排序，而有些场景不需要排序，Spark 可以避免不必要的排序所带来的开销。
- Spark 是一张有向无环图（从一个点出发最终无法回到该点的一个拓扑），并对其进行优化。

4. Spark 支持的 API

Spark 支持的 API 包括 Scala、Python、Java 等。

5. 运行模式

- Local（用于测试、开发）。
- Standlone（独立集群模式）。
- Spark on YARN（Spark 在 YARN 上）。
- Spark on Mesos（Spark 在 Mesos 上）。

6. Spark 运行时的步骤

Driver 程序启动多个 Worker，Worker 从文件系统加载数据并产生 RDD（即数据放到 RDD 中，RDD 是一个数据结构），并按照不同分区 Cache 到内存中。Spark 执行步骤如图 11-2 所示。

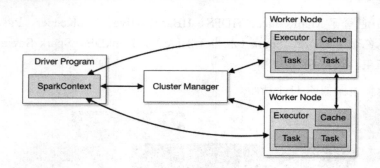

图 11-2　Spark 执行步骤

7. RDD

RDD 英文名为 Resilient Distributed Dataset，中文名为弹性分布式数据集。

什么是 RDD？RDD 是一个只读、分区记录的集合，你可以把它理解为一个存储数据的数据结构！在 Spark 中一切操作基于 RDD。

RDD 可以通过以下几种方式创建：

- 集合转换。
- 从文件系统（本地文件、HDFS、HBase）输入。
- 从父 RDD 转换（为什么需要父 RDD 呢？容错，下面会提及）。

RDD 的计算类型：

- Transformation：延迟执行，一个 RDD 通过该操作产生新的 RDD 时不会立即执行，只有等到 Action 操作才会真正执行。
- Action：提交 Spark 作业，当 Action 时，Transformation 类型的操作才会真正执行计算操作，然后产生最终结果输出。
- Hadoop 提供处理的数据接口有 Map 和 Reduce，而 Spark 提供的不仅仅有 Map 和 Reduce，还有更多对数据处理的接口。Spark 算子如图 11-3 所示。

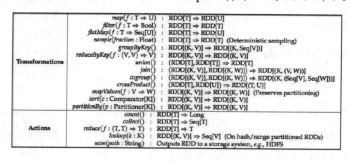

图 11-3　Spark 算子

8. 容错 Lineage

每个 RDD 都会记录自己所依赖的父 RDD，一旦出现某个 RDD 的某些 Partition 丢失，可以通过并行计算迅速恢复，这就是容错。

RDD 的依赖又分为 Narrow Dependent（窄依赖）和 Wide Dependent（宽依赖）。

（1）窄依赖：每个 Partition 最多只能给一个 RDD 使用，由于没有多重依赖，所以在一个节点上可以一次性将 Partition 处理完，且一旦数据发生丢失或者损坏，可以迅速从上一个 RDD 恢复。

（2）宽依赖：每个 Partition 可以给多个 RDD 使用，由于多重依赖，只有等到所有到达节点的数据处理完毕才能进行下一步处理，一旦发生数据丢失或者损坏，则完蛋了，所以在此发生之前，必须将上一次所有节点的数据进行物化（存储到磁盘上）处理，这样达到恢复。

RDD 宽、窄依赖如图 11-4 所示。

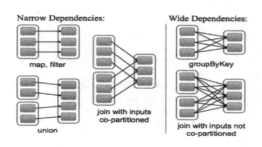

图 11-4　RDD 宽、窄依赖

9. 缓存策略

Spark 通过 useDisk、useMemory、deserialized、replication 4 个参数组成 11 种缓存策略。

- useDisk：使用磁盘缓存（boolean）。
- useMemory：使用内存缓存（boolean）。
- deserialized：反序列化（序列化是为了网络将对象进行传输，boolean：true 反序列化\false 序列化）。
- replication：副本数量（int）。

通过 StorageLevel 类的构造传参的方式进行控制，结构如下：

```
class StorageLevel private(useDisk : Boolean ,useMemory :
Boolean ,deserialized : Boolean ,replication: Ini)
```

10. 提交的方式

- spark-submit（官方推荐）
- sbt run
- ava -jar

提交时可以指定各种参数。

```
./bin/spark-submit
-- class  <main- class >
```

```
--master <master-url>
--deploy-mode <deploy-mode>
--conf <key> = <value>
...  # other options
<application-jar>
[application-arguments]
```

spark-submit 提交方式如图 11-5 所示。

```
# Run application locally on 8 cores
./bin/spark-submit \
  --class org.apache.spark.examples.SparkPi \
  --master local[8] \
  /path/to/examples.jar \
  100

# Run on a Spark Standalone cluster in client deploy mode
./bin/spark-submit \
  --class org.apache.spark.examples.SparkPi \
  --master spark://207.184.161.138:7077 \
  --executor-memory 20G \
  --total-executor-cores 100 \
  /path/to/examples.jar \
  1000
```

图 11-5　spark-submit 提交方式

11.3 Spark 算子实战及功能描述

Spark 算子大致上可分三大类算子：

- Value 数据类型的 Transformation 算子，这种变换不触发提交作业，针对处理的数据项是 Value 型的数据。
- Key-Value 数据类型的 Transformation 算子，这种变换不触发提交作业，针对处理的数据项是 Key-Value 型的数据。
- Action 算子，这类算子会触发 SparkContext 提交作业。

11.3.1　Value 型 Transformation 算子

1. map

数据集中的每个元素经过用户自定义的函数转换形成一个新的 RDD，新的 RDD 叫 MappedRDD。

```
val a = sc.parallelize(List("dog", "salmon", "salmon", "rat", "elephant"), 3)
val b = a.map(_.length)
```

```
val c = a.zip(b)
c.collect
```

zip 函数用于将两个 RDD 组合成 Key/Value 形式的 RDD。

结果：

res0: Array[(String, Int)] = Array((dog,3), (salmon,6), (salmon,6), (rat,3), (elephant,8))

2. flatMap

描述：与 map 类似，但每个元素输入项都可以被映射到 0 个或多个的输出项，最终将结果"扁平化"后输出。

```
val a = sc.parallelize(1 to 10, 5)
a.flatMap(1 to   ).collect
```

结果：

res1: Array[Int] = Array(1, 1, 2, 1, 2, 3, 1, 2, 3, 4, 1, 2, 3, 4, 5, 1, 2, 3, 4, 5, 6, 1, 2, 3, 4, 5, 6, 7, 1, 2, 3, 4, 5, 6, 7, 8, 1, 2, 3, 4, 5, 6, 7, 8, 9, 1, 2, 3, 4, 5, 6, 7, 8, 9, 10)

```
sc.parallelize(List(1, 2, 3), 2).flatMap(x => List(x, x, x)).collect
```

结果： res2: Array[Int] = Array(1, 1, 1, 2, 2, 2, 3, 3, 3)

3. mapPartitions

描述：类似于 map，map 作用于每个分区的每个元素，但 mapPartitions 作用于每个分区的 func 的类型：Iterator[T] => Iterator[U] 假设有 N 个元素，有 M 个分区，那么 map 的函数的将被调用 N 次，而 mapPartitions 被调用 M 次，当在映射的过程中不断地创建对象时就可以使用 mapPartitions，比 map 的效率要高很多。比如：当向数据库写入数据时，如果使用 map，就需要为每个元素创建 connection 对象；但使用 mapPartitions 的话，就需要为每个分区创建 connection 对象。

```
val l =
List(("kpop","female"),("zorro","male"),("mobin","male"),("lucy","female"))
val rdd = sc.parallelize(l,2)
rdd.mapPartitions(x => x.filter(_._2 ==
"female")).foreachPartition(p=>{
println(p.toList)
   println("====分区分割线====" )
})
```

结果：

====分区分割线====

List((kpop,female))

====分区分割线====

List((lucy,female))

4. glom

将 RDD 的每个分区中的类型为 T 的元素转换为数组 Array[T]。

```
val a = sc.parallelize(1 to 100, 3)
a.glom.collect
```

结果：

res3: Array[Array[Int]] = Array(Array(1, 2, 3, 4, 5, 6, 7, 8, 9, 10, 11, 12, 13, 14, 15, 16, 17, 18, 19, 20, 21, 22, 23, 24, 25, 26, 27, 28, 29, 30, 31, 32, 33), Array(34, 35, 36, 37, 38, 39, 40, 41, 42, 43, 44, 45, 46, 47, 48, 49, 50, 51, 52, 53, 54, 55, 56, 57, 58, 59, 60, 61, 62, 63, 64, 65, 66), Array(67, 68, 69, 70, 71, 72, 73, 74, 75, 76, 77, 78, 79, 80, 81, 82, 83, 84, 85, 86, 87, 88, 89, 90, 91, 92, 93, 94, 95, 96, 97, 98, 99, 100))

5. union

UNION 指将两个 RDD 中的数据集进行合并，最终返回两个 RDD 的并集，若 RDD 中存在相同的元素，也不会去重。

```
val a = sc.parallelize(1 to 3, 1)
val b = sc.parallelize(1 to 7, 1)
(a ++ b).collect
```

结果：

res4: Array[Int] = Array(1, 2, 3, 5, 6, 7)

6. cartesian

对两个 RDD 中的所有元素进行笛卡尔积操作。

```
val x = sc.parallelize(List(1,2,3,4,5))
val y = sc.parallelize(List(6,7,8,9,10))
x.cartesian(y).collect
```

结果：

res5: Array[(Int, Int)] = Array((1,6), (1,7), (1,8), (1,9), (1,10), (2,6), (2,7), (2,8), (2,9), (2,10), (3,6), (3,7), (3,8), (3,9), (3,10), (4,6), (5,6), (4,7), (5,7), (4,8), (5,8), (4,9), (4,10), (5,9), (5,10))

7. groupBy

生成相应的 key，相同的放在一起 even（2,4,6,8）。

```
val a = sc.parallelize(1 to 9, 3)
a.groupBy(x => { if (x % 2 == 0) "even" else "odd" }).collect
```

结果：

res6: Array[(String, Seq[Int])] = Array((even,ArrayBuffer(2, 4, 6, 8)), (odd,ArrayBuffer(1, 3, 5, 7, 9)))

8. filter

对元素进行过滤，对每个元素应用 f 函数，返回值为 true 的元素在 RDD 中保留，返回为 false 的将过滤掉。

```
val a = sc.parallelize(1 to 10, 3)
val b = a.filter(_ % 2 == 0)
b.collect
```

结果：

res7: Array[Int] = Array(2, 4, 6, 8, 10)

9. distinct

distinct 用于去重。

```
val c = sc.parallelize(List("Gnu", "Cat", "Rat", "Dog", "Gnu",
"Rat"), 2)
c.distinct.collect
```

结果：

res8: Array[String] = Array(Dog, Gnu, Cat, Rat)

10. subtract

去掉含有重复的项。

```
val a = sc.parallelize(1 to 9, 3)
val b = sc.parallelize(1 to 3, 3)
val c = a.subtract(b)
c.collect
```

结果：

res9: Array[Int] = Array(6, 9, 4, 7, 5, 8)

11. sample

以指定的随机种子随机抽样出数量为 fraction 的数据，withReplacement 表示是抽出的数据是否放回，true 为有放回的抽样，false 为无放回的抽样。

```
val a = sc.parallelize(1 to 10000, 3)
a.sample(false, 0.1, 0).count
```

结果：

res10: Long = 999

12. takesample

takeSample()函数和 sample 函数是一个原理，但是不使用相对比例采样，而是按设定的采样个数进行采样，同时返回结果不再是 RDD，而是相当于对采样后的数据进行 collect()，返回结果的集合为单机的数组。

```
val x = sc.parallelize(1 to 1000, 3)
x.takeSample(true, 100, 1)
```

结果：

res11: Array[Int] = Array(339, 718, 810, 105, 71, 268, 333, 360, 341, 300, 68, 848, 431, 449, 773, 172, 802, 339, 431, 285, 937, 301, 167, 69, 330, 864, 40, 645, 65, 349, 613, 468, 982, 314, 160,

675, 232, 794, 577, 571, 805, 317, 136, 860, 522, 45, 628, 178, 321, 482, 657, 114, 332, 728, 901, 290, 175, 876, 227, 130, 863, 773, 559, 301, 694, 460, 839, 952, 664, 851, 260, 729, 823, 880, 792, 964, 614, 821, 683, 364, 80, 875, 813, 951, 663, 344, 546, 918, 436, 451, 397, 670, 756, 512, 391, 70, 213, 896, 123, 858)

13. cache、persist

cache 和 persist 都是用于将一个 RDD 进行缓存的，这样在之后使用的过程中就不需要重新计算了，可以大大节省程序运行时间。

```
val c = sc.parallelize(List("Gnu", "Cat", "Rat", "Dog", "Gnu", "Rat"), 2)
c.getStorageLevel
```

结果：

res12: org.apache.spark.storage.StorageLevel = StorageLevel(false, false, false, false, 1)

```
c.cache
c.getStorageLevel
```

结果：

res13: org.apache.spark.storage.StorageLevel = StorageLevel(false, true, false, true, 1)

11.3.2 Key-Value 型 Transformation 算子

1. mapValues

mapValues 是针对[K,V]中的 V 值进行 map 操作。

```
val a = sc.parallelize(List("dog", "tiger", "lion", "cat", "panther", "eagle"), 2)
val b = a.map(x => (x.length, x))
b.mapValues("x" +  + "x").collect
```

结果：

res14: Array[(Int, String)] = Array((3,xdogx), (5,xtigerx), (4,xlionx), (3,xcatx), (7,xpantherx), (5,xeaglex))

2. combineByKey

使用用户设置好的聚合函数对每个 Key 中的 Value 进行组合（combine），可以将输入类型为 RDD[(K, V)]转成 RDD[(K, C)]。

```
val a = sc.parallelize(List("dog","cat","gnu","salmon","rabbit","turkey","wolf","bear","bee"), 3)
val b = sc.parallelize(List(1,1,2,2,2,1,2,2,2), 3)
val c = b.zip(a)
val d = c.combineByKey(List(_), (x:List[String], y:String) => y :: x, (x:List[String], y:List[String]) => x ::: y)
```

```
d.collect
```

结果：

res15: Array[(Int, List[String])] = Array((1,List(cat, dog, turkey)), (2,List(gnu, rabbit, salmon, bee, bear, wolf)))

3. reduceByKey

对元素为 KV 对的 RDD 中 Key 相同的元素的 Value 进行 binary_function 的 reduce 操作，因此，Key 相同的多个元素的值被 reduce 为一个值，然后与原 RDD 中的 Key 组成一个新的 KV 对。

```
val a = sc.parallelize(List("dog", "cat", "owl", "gnu", "ant"), 2)
val b = a.map(x => (x.length, x))
b.reduceByKey( _ + _ ).collect
```

结果：

res16: Array[(Int, String)] = Array((3,dogcatowlgnuant))

```
val a = sc.parallelize(List("dog", "tiger", "lion", "cat", "panther", "eagle"), 2)
val b = a.map(x => (x.length, x))
b.reduceByKey( _ + _ ).collect
```

结果：

res17: Array[(Int, String)] = Array((4,lion), (3,dogcat), (7,panther), (5,tigereagle))

4. partitionBy

对 RDD 进行分区操作。

5. cogroup

cogroup 指对两个 RDD 中的 KV 元素，每个 RDD 中相同 key 中的元素分别聚合成一个集合。

```
val a = sc.parallelize(List(1, 2, 1, 3), 1)
val b = a.map((_, "b"))
val c = a.map((_, "c"))
b.cogroup(c).collect
```

结果：

res18: Array[(Int, (Iterable[String], Iterable[String]))] = Array(

(2,(ArrayBuffer(b),ArrayBuffer(c))),

(3,(ArrayBuffer(b),ArrayBuffer(c))),

(1,(ArrayBuffer(b, b),ArrayBuffer(c, c))))

6. join

对两个需要连接的 RDD 进行 cogroup 函数操作。

```
val a = sc.parallelize(List("dog", "salmon", "salmon", "rat", "elephant"), 3)
```

```
val b = a.keyBy(_.length)
val c =
sc.parallelize(List("dog","cat","gnu","salmon","rabbit","turkey",
"wolf","bear","bee"), 3)
val d = c.keyBy(_.length)
b.join(d).collect
```

结果：

res19: Array[(Int, (String, String))] = Array((6,(salmon,salmon)), (6,(salmon,rabbit)), (6,(salmon,turkey)), (6,(salmon,salmon)), (6,(salmon,rabbit)), (6,(salmon,turkey)), (3,(dog,dog)), (3,(dog,cat)), (3,(dog,gnu)), (3,(dog,bee)), (3,(rat,dog)), (3,(rat,cat)), (3,(rat,gnu)), (3,(rat,bee)))

7. leftOutJoin

```
val a = sc.parallelize(List("dog", "salmon", "salmon", "rat",
"elephant"), 3)
val b = a.keyBy(_.length)
val c =
sc.parallelize(List("dog","cat","gnu","salmon","rabbit","turkey","wo
lf","bear","bee"), 3)
val d = c.keyBy(_.length)
b.leftOuterJoin(d).collect
```

结果：

res20: Array[(Int, (String, Option[String]))] = Array((6,(salmon,Some(salmon))), (6,(salmon,Some(rabbit))), (6,(salmon,Some(turkey))), (6,(salmon,Some(salmon))), (6,(salmon,Some(rabbit))), (6,(salmon,Some(turkey))), (3,(dog,Some(dog))), (3,(dog,Some(cat))), (3,(dog,Some(gnu))), (3,(dog,Some(bee))), (3,(rat,Some(dog))), (3,(rat,Some(cat))), (3,(rat,Some(gnu))), (3,(rat,Some(bee))), (8,(elephant,None)))

8. rightOutJoin

```
val a = sc.parallelize(List("dog", "salmon", "salmon", "rat",
"elephant"), 3)
val b = a.keyBy(_.length)
val c =
sc.parallelize(List("dog","cat","gnu","salmon","rabbit","turkey",
"wolf","bear","bee"), 3)
val d = c.keyBy(_.length)
b.rightOuterJoin(d).collect
```

结果：

res21: Array[(Int, (Option[String], String))] = Array((6,(Some(salmon),salmon)), (6,(Some(salmon),rabbit)), (6,(Some(salmon),turkey)), (6,(Some(salmon),salmon)), (6,(Some(salmon),rabbit)), (6,(Some(salmon),turkey)), (3,(Some(dog),dog)), (3,(Some(dog),cat)), (3,(Some(dog),gnu)), (3,(Some(dog),bee)), (3,(Some(rat),dog)), (3,(Some(rat),cat)), (3,(Some(rat),gnu)), (3,(Some(rat),bee)), (4,(None,wolf)), (4,(None,bear)))

11.3.3 Actions 算子

1. foreach

描述：打印输出。

```
val c = sc.parallelize(List("cat", "dog", "tiger", "lion", "gnu",
"crocodile", "ant", "whale", "dolphin", "spider"), 3)
c.foreach(x => println(x + "s are yummy"))
```

结果：

lions are yummy

gnus are yummy

crocodiles are yummy

ants are yummy

whales are yummy

dolphins are yummy

spiders are yummy

2. saveAsTextFile

保存结果到 HDFS。

```
val a = sc.parallelize(1 to 10000, 3)
a.saveAsTextFile("/user/yuhui/mydata_a")
```

结果：

[root@tagtic-slave03 ~]# Hadoop fs -ls /user/yuhui/mydata_a

Found 4 items

-rw-r–r– 2 root supergroup 0 2017-05-22 14:28 /user/yuhui/mydata_a/_SUCCESS

-rw-r–r– 2 root supergroup 15558 2017-05-22 14:28 /user/yuhui/mydata_a/part-00000

-rw-r–r– 2 root supergroup 16665 2017-05-22 14:28 /user/yuhui/mydata_a/part-00001

-rw-r–r– 2 root supergroup 16671 2017-05-22 14:28 /user/yuhui/mydata_a/part-00002

3. saveAsObjectFile

saveAsObjectFile 用于将 RDD 中的元素序列化成对象，存储到文件中。对于 HDFS，默认采用 SequenceFile 保存。

```
val x = sc.parallelize(1 to 100, 3)
x.saveAsObjectFile("/user/yuhui/objFile")
val y = sc.objectFile[Int]("/user/yuhui/objFile")
y.collect
```

结果：

res22: Array[Int] = Array[Int] = Array(1, 2, 3, 4, 5, 6, 7, 8, 9, 10, 11, 12, 13, 14, 15, 16, 17, 18, 19, 20, 21, 22, 23, 24, 25, 26, 27, 28, 29, 30, 31, 32, 33, 34, 35, 36, 37, 38, 39, 40, 41, 42, 43, 44, 45, 46, 47, 48, 49, 50, 51, 52, 53, 54, 55, 56, 57, 58, 59, 60, 61, 62, 63, 64, 65, 66, 67, 68, 69, 70, 71, 72, 73, 74, 75, 76, 77, 78, 79, 80, 81, 82, 83, 84, 85, 86, 87, 88, 89, 90, 91, 92, 93, 94, 95, 96, 97, 98, 99, 100)

4. collect

将 RDD 中的数据收集起来，变成一个 Array，仅限数据量比较小的时候。

```
val c = sc.parallelize(List("Gnu", "Cat", "Rat", "Dog", "Gnu", "Rat"), 2)
c.collect
```

结果：

res23: Array[String] = Array(Gnu, Cat, Rat, Dog, Gnu, Rat)

5. collectAsMap

返回 hashMap 包含所有 RDD 中的分片，key 如果重复，后边的元素会覆盖前面的元素。
zip 函数用于将两个 RDD 组合成 Key/Value 形式的 RDD。

```
val a = sc.parallelize(List(1, 2, 1, 3), 1)
val b = a.zip(a)
b.collectAsMap
```

结果：

res24: Scala.collection.Map[Int,Int] = Map(2 -> 2, 1 -> 1, 3 -> 3)

6. reduceByKeyLocally

先执行 reduce，然后再执行 collectAsMap。

```
val a = sc.parallelize(List("dog", "cat", "owl", "gnu", "ant"), 2)
val b = a.map(x => (x.length, x))
b.reduceByKey(_ + _).collect
```

结果：

res25: Array[(Int, String)] = Array((3,dogcatowlgnuant))

7. lookup

查找，针对 key-value 类型的 RDD。

```
val a = sc.parallelize(List("dog", "tiger", "lion", "cat", "panther", "eagle"), 2)
val b = a.map(x => (x.length, x))
b.lookup(3)
```

结果：

res26: Seq[String] = WrappedArray(tiger, eagle)

8. count

总数。

```
val c = sc.parallelize(List("Gnu", "Cat", "Rat", "Dog"), 2)
c.count
```

结果：

res27: Long = 4

9. top

返回最大的 K 个元素。

```
val c = sc.parallelize(Array(6, 9, 4, 7, 5, 8), 2)
c.top(2)
```

结果：

res28: Array[Int] = Array(9, 8)

10. reduce

相当于对 RDD 中的元素进行 reduceLeft 函数的操作。

```
val a = sc.parallelize(1 to 100, 3)
a.reduce(_ + _)
```

结果：

res29: Int = 5050

11. fold

fold()与 reduce()类似，接收与 reduce 接收的函数签名相同的函数，另外再加上一个初始值作为第一次调用的结果。结果为：（区+1）*（初始值）+list（值）。

```
val a = sc.parallelize(List(1,2,3), 3)
a.fold(0)(_ + _)
```

结果：

res30: Int = 6

12. aggregate

aggregate 先对每个分区的所有元素进行 aggregate 操作，再对分区的结果进行 fold 操作。

```
val z = sc.parallelize(List(1,2,3,4,5,6), 2)

// lets first print out the contents of the RDD with partition labels
def myfunc(index: Int, iter: Iterator[(Int)]) : Iterator[String] = {
  iter.toList.map(x => "[partID:" + index + ", val: " + x + "]").iterator
}

z.mapPartitionsWithIndex(myfunc).collect
```

结果：

res31: Array[String] = Array([partID:0, val: 1], [partID:0, val: 2], [partID:0, val: 3], [partID:1, val: 4], [partID:1, val: 5], [partID:1, val: 6])

```
z.aggregate(0)(math.max(_ , _), _ + _)
```

结果:

res32: Int = 9

11.4 Spark Streaming 实战

1. Spark Streaming 是什么

Spark Streaming 是核心 Spark API 的扩展,它支持可伸缩、高吞吐量、可容错地处理实时数据流。数据可以从 Kafka、Flume、Twitter、ZeroMQ、Kinesis、TCP sockets 等许多资源中获取,可以使用复杂的算法来处理,如 map、reduce、join 和 window 等高级函数。最后,可以将处理过的数据推送到文件系统、数据库和实时仪表板上。实际上,你可以在数据流上应用 Spark 的机器学习和图形处理算法。Spark Streaming 部署选项如图 11-6 所示。

图 11-6　Spark Streaming 部署选项

2. Spark Streaming 的 A Quick Example

A Quick Example 测试代码如下,解说如图 11-7 所示。

```
packagecn.donews.bi

/**
 * Created by yuhui on 2017/3/27.
 */

import org.apache.spark._
import org.apache.spark.streaming._

object Demo01{
defmain(args: Array[String]): Unit = {
valconf = new
SparkConf().setMaster("local[2]").setAppName("NetworkWordCount")
valssc = new StreamingContext(conf, Seconds(1))

vallines = ssc.socketTextStream("localhost", 9999)

valwords = lines.flatMap(_.split(" "))

// Count each word in each batch
```

```
valpairs = words.map(word => (word, 1))
valwordCounts = pairs.reduceByKey(_ + _)

// Print the first ten elements of each RDD generated in this DStream
to the console
wordCounts.print()

ssc.start()             // Start the computation
ssc.awaitTermination()  // Wait for the computation to terminate

}
}
```

图 11-7　测试代码解说

A Quick Example 测试结果展示如图 11-8 所示。

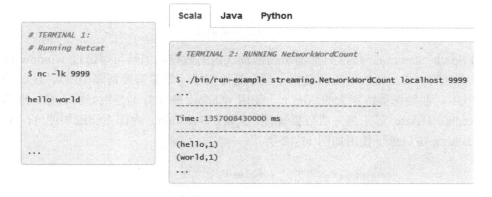

图 11-8　测试结果展示

```xml
<dependency>
<groupId>org.apache.spark</groupId>
<artifactId>spark-streaming_2.11</artifactId>
<version>2.1.0</version>
</dependency>
```

3. Discretized Streams (DStreams)

离散流或 DStream 是由 Spark Streaming 提供的基本抽象。DStream 由一组连续的 RDDs 表示。DStream 与 RDD 关系如图 11-9 所示。

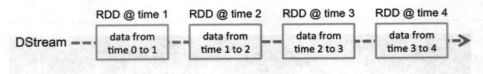

图 11-9　DStream 与 RDD 关系

上图中，Dstream 就是一个基础抽象的管道，每一个 Duration 就是一个 RDD。

4. Dstream 时间窗口

Spark Streaming 还提供了窗口计算，允许你在数据的滑动窗口上应用转换。DStream 时间窗口如图 11-10 所示，图中演示了这个滑动窗口。

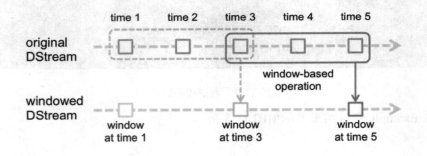

图 11-10　Dstream 时间窗口

使用 Spark Streaming 每次只能消费当前批次内的数据，当然可以通过 window 操作，消费过去一段时间（多个批次）内的数据。举个简单例子，如果需要每隔 10 秒，统计当前小时的 PV 和 UV，在数据量特别大的情况下，使用 window 操作并不是很好的选择，通常是借助其他如 Redis、HBase 等工具完成数据统计。Spark Streaming 内部工作流如图 11-11 所示。Spark Streaming 窗口算子使用如图 11-12 所示。

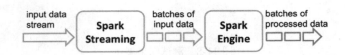

图 11-11　Spark Streaming 内部工作流程

```
object KafkaWordCount {
  def main(args: Array[String]) {
    if (args.length < 4) {
      System.err.println("Usage: KafkaWordCount <zkQuorum> <group> <topics> <numThreads>")
      System.exit(1)
    }

    StreamingExamples.setStreamingLogLevels()

    val Array(zkQuorum, group, topics, numThreads) = args
    val sparkConf = new SparkConf().setAppName("KafkaWordCount")
    val ssc = new StreamingContext(sparkConf, Seconds(2))
    ssc.checkpoint("checkpoint")

    val topicMap = topics.split(",").map((_, numThreads.toInt)).toMap
    val lines = KafkaUtils.createStream(ssc, zkQuorum, group, topicMap).map(_._2)
    val words = lines.flatMap(_.split(" "))
    val wordCounts = words.map(x => (x, 1L))
      .reduceByKeyAndWindow(_ + _, _ - _, Minutes(10), Seconds(2), 2)
    wordCounts.print()

    ssc.start()
    ssc.awaitTermination()
  }
}
```

图 11-12 Spark Streaming 窗口算子使用

注意：时间窗口的时间一定是每个 Duration 产生成 RDD 的倍数。

5. Dstream 操作

Duration 是用户自定义的时间间隔，每个间隔会产生一个 RDD。Dstream 和 Duration 关系如图 11-13 所示。

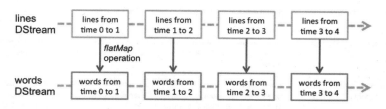

图 11-13 Dstream 和 Duration 关系

（1）转换操作（Transformations on Dstreams），转换操作算子如图 11-14 所示。

Transformations on DStreams

Similar to that of RDDs, transformations allow the data from the input DStream to be modified. DStreams support many of the transformations available on normal Spark RDD's. Some of the common ones are as follows.

Transformation	Meaning
map(*func*)	Return a new DStream by passing each element of the source DStream through a function *func*.
flatMap(*func*)	Similar to map, but each input item can be mapped to 0 or more output items.
filter(*func*)	Return a new DStream by selecting only the records of the source DStream on which *func* returns true.
repartition(*numPartitions*)	Changes the level of parallelism in this DStream by creating more or fewer partitions.
union(*otherStream*)	Return a new DStream that contains the union of the elements in the source DStream and *otherDStream*.
count()	Return a new DStream of single-element RDDs by counting the number of elements in each RDD of the source DStream.
reduce(*func*)	Return a new DStream of single-element RDDs by aggregating the elements in each RDD of the source DStream using a function *func* (which takes two arguments and returns one). The function should be associative so that it can be computed in parallel.
countByValue()	When called on a DStream of elements of type K, return a new DStream of (K, Long) pairs where the value of each key is its frequency in each RDD of the source DStream.

reduceByKey(*func*, [*numTasks*])	When called on a DStream of (K, V) pairs, return a new DStream of (K, V) pairs where the values for each key are aggregated using the given reduce function. **Note**: By default, this uses Spark's default number of parallel tasks (2 for local mode, and in cluster mode the number is determined by the config property `spark.default.parallelism`) to do the grouping. You can pass an optional `numTasks` argument to set a different number of tasks.
join(*otherStream*, [*numTasks*])	When called on two DStreams of (K, V) and (K, W) pairs, return a new DStream of (K, (V, W)) pairs with all pairs of elements for each key.
cogroup(*otherStream*, [*numTasks*])	When called on a DStream of (K, V) and (K, W) pairs, return a new DStream of (K, Seq[V], Seq[W]) tuples.
transform(*func*)	Return a new DStream by applying a RDD-to-RDD function to every RDD of the source DStream. This can be used to do arbitrary RDD operations on the DStream.
updateStateByKey(*func*)	Return a new "state" DStream where the state for each key is updated by applying the given function on the previous state of the key and the new values for the key. This can be used to maintain arbitrary state data for each key.

图 11-14 转换操作算子

（2）基于窗口的转换操作（Window Operations），基于窗口的转换操作算子如图 11-15 所示。

Transformation	Meaning
window(*windowLength*, *slideInterval*)	Return a new DStream which is computed based on windowed batches of the source DStream.
countByWindow(*windowLength*, *slideInterval*)	Return a sliding window count of elements in the stream.
reduceByWindow(*func*, *windowLength*, *slideInterval*)	Return a new single-element stream, created by aggregating elements in the stream over a sliding interval using *func*. The function should be associative so that it can be computed correctly in parallel.
reduceByKeyAndWindow(*func*, *windowLength*, *slideInterval*, [*numTasks*])	When called on a DStream of (K, V) pairs, returns a new DStream of (K, V) pairs where the values for each key are aggregated using the given reduce function *func* over batches in a sliding window. **Note**: By default, this uses Spark's default number of parallel tasks (2 for local mode, and in cluster mode the number is determined by the config property `spark.default.parallelism`) to do the grouping. You can pass an optional `numTasks` argument to set a different number of tasks.
reduceByKeyAndWindow(*func*, *invFunc*, *windowLength*, *slideInterval*, [*numTasks*])	A more efficient version of the above `reduceByKeyAndWindow()` where the reduce value of each window is calculated incrementally using the reduce values of the previous window. This is done by reducing the new data that enters the sliding window, and "inverse reducing" the old data that leaves the window. An example would be that of "adding" and "subtracting" counts of keys as the window slides. However, it is applicable only to "invertible reduce functions", that is, those reduce functions which have a corresponding "inverse reduce" function (taken as parameter *invFunc*). Like in `reduceByKeyAndWindow`, the number of reduce tasks is configurable through an optional argument. Note that checkpointing must be enabled for using this operation.
countByValueAndWindow(*windowLength*, *slideInterval*, [*numTasks*])	When called on a DStream of (K, V) pairs, returns a new DStream of (K, Long) pairs where the value of each key is its frequency within a sliding window. Like in `reduceByKeyAndWindow`, the number of reduce tasks is configurable through an optional argument.

图 11-15 基于窗口的转换操作算子

（3）输出操作（Output Operations on DStreams），输出算子如图 11-16 所示。

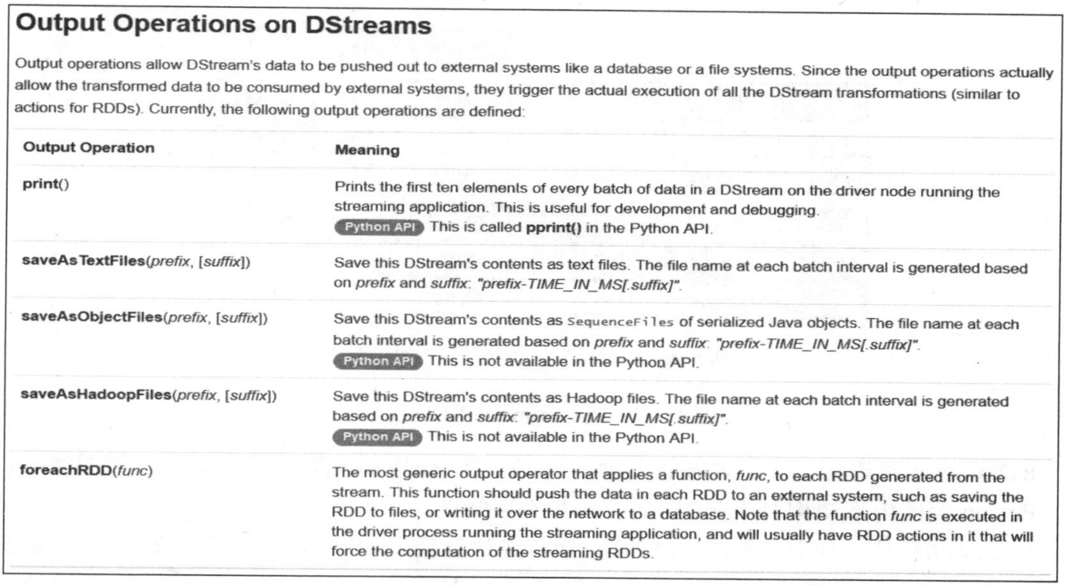

```
Output Operations on DStreams
Output operations allow DStream's data to be pushed out to external systems like a database or a file systems. Since the output operations actually
allow the transformed data to be consumed by external systems, they trigger the actual execution of all the DStream transformations (similar to
actions for RDDs). Currently, the following output operations are defined:

Output Operation               Meaning
print()                        Prints the first ten elements of every batch of data in a DStream on the driver node running the
                               streaming application. This is useful for development and debugging.
                               Python API  This is called pprint() in the Python API.
saveAsTextFiles(prefix, [suffix])   Save this DStream's contents as text files. The file name at each batch interval is generated based
                               on prefix and suffix: "prefix-TIME_IN_MS[.suffix]".
saveAsObjectFiles(prefix, [suffix]) Save this DStream's contents as SequenceFiles of serialized Java objects. The file name at each
                               batch interval is generated based on prefix and suffix: "prefix-TIME_IN_MS[.suffix]".
                               Python API  This is not available in the Python API.
saveAsHadoopFiles(prefix, [suffix]) Save this DStream's contents as Hadoop files. The file name at each batch interval is generated
                               based on prefix and suffix: "prefix-TIME_IN_MS[.suffix]".
                               Python API  This is not available in the Python API.
foreachRDD(func)               The most generic output operator that applies a function, func, to each RDD generated from the
                               stream. This function should push the data in each RDD to an external system, such as saving the
                               RDD to files, or writing it over the network to a database. Note that the function func is executed in
                               the driver process running the streaming application, and will usually have RDD actions in it that will
                               force the computation of the streaming RDDs.
```

图 11-16　输出算子

11.5　Spark SQL 和 DataFrame 实战

1. Spark SQL 和 DataFrame 概念

（1）什么是 Spark SQL

Spark SQL 的一个用途是执行使用基本 SQL 语法或 HiveQL 编写的 SQL 查询。Spark SQL 还可以用于从现有的 Hive 安装中读取数据。当从另一种编程语言中运行 SQL 时，结果将作为 DataFrame 返回。你还可以使用命令行或 jdbc/odbc 来与 SQL 接口进行交互。

（2）什么是 DataFrame

一个以命名列组织的分布式数据集。概念上相当于关系数据库中一张表或在 R / Python 中的 DataFrame 数据结构，但 DataFrame 有丰富的优化。在 Spark 1.3 之前，核心的新类型为 RDD-schemaRDD，后面改为 DataFrame。Spark 通过 DataFrame 操作大量的数据源，包括外部文件（如 json、avro、parquet、sequencefile 等）、Hive、关系数据库、Cassandra 等。

（3）DataFrame 和 RDD 的区别

DataFrame 和 RDD 的区别如图 11-17 所示。

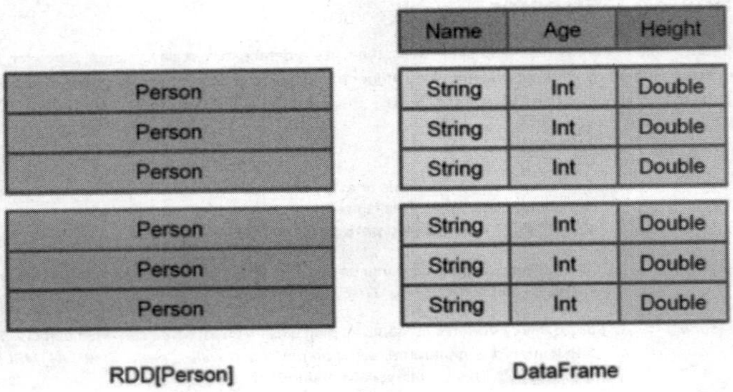

Name	Age	Height
String	Int	Double
String	Int	Double
String	Int	Double
String	Int	Double
String	Int	Double
String	Int	Double

RDD[Person]　　　　　　　　DataFrame

图 11-17　DataFrame 与 RDD 的区别

RDD 是分布式的 Java 对象的集合，比如，RDD[Person]是以 Person 为类型参数，但是，Person 类的内部结构对于 RDD 而言却是不可知的。

DataFrame 是一种以 RDD 为基础的分布式数据集，也就是分布式的 Row 对象的集合（每个 Row 对象代表一行记录），提供了详细的结构信息，也就是我们经常说的模式（schema），Spark SQL 可以清楚地知道该数据集中包含哪些列、每列的名称和类型。

和 RDD 一样，DataFrame 的各种变换操作也采用惰性机制，只是记录了各种转换的逻辑转换路线图（是一个 DAG 图），不会发生真正的计算，这个 DAG 图相当于一个逻辑查询计划，最终，会被翻译成物理查询计划，生成 RDD DAG，按照之前介绍的 RDD DAG 的执行方式去完成最终的计算，最后得到结果。

2. 数据准备

数据文件内容如表 11-1 所示。

表 11-1　数据文件内容

people.json	people.txt
{"name":"Michael"}	Michael, 29
{"name":"Andy", "age":30}	Andy, 30
{"name":"Justin", "age":19}	Justin, 19

3. spark-shell 启动及参数

```
[root@tagtic-master sql]# spark-shell --driver-memory 1G --executor-memory 1G --executor-cores 4 --name yuhui
```

默认初始化：

```
Spark context available as sc.
SQL context available as sqlContext.
```

spark-shell 启动及参数可以通过执行命令 spark-shell--help 查看，如图 11-18、图 11-19 所示。

```
[root@tagtic-master sql]# spark-shell --help
Usage: ./bin/spark-shell [options]

Options:
  --master MASTER_URL         spark://host:port, mesos://host:port, yarn, or local.
  --deploy-mode DEPLOY_MODE   Whether to launch the driver program locally ("client") or
                              on one of the worker machines inside the cluster ("cluster")
                              (Default: client).
  --class CLASS_NAME          Your application's main class (for Java / Scala apps).
  --name NAME                 A name of your application.
  --jars JARS                 Comma-separated list of local jars to include on the driver
                              and executor classpaths.
  --packages                  Comma-separated list of maven coordinates of jars to include
                              on the driver and executor classpaths. Will search the local
                              maven repo, then maven central and any additional remote
                              repositories given by --repositories. The format for the
                              coordinates should be groupId:artifactId:version.
  --exclude-packages          Comma-separated list of groupId:artifactId, to exclude while
                              resolving the dependencies provided in --packages to avoid
                              dependency conflicts.
  --repositories              Comma-separated list of additional remote repositories to
                              search for the maven coordinates given with --packages.
  --py-files PY_FILES         Comma-separated list of .zip, .egg, or .py files to place
                              on the PYTHONPATH for Python apps.
  --files FILES               Comma-separated list of files to be placed in the working
                              directory of each executor.

  --conf PROP=VALUE           Arbitrary Spark configuration property.
  --properties-file FILE      Path to a file from which to load extra properties. If not
                              specified, this will look for conf/spark-defaults.conf.

  --driver-memory MEM         Memory for driver (e.g. 1000M, 2G) (Default: 1024M).
  --driver-java-options       Extra Java options to pass to the driver.
  --driver-library-path       Extra library path entries to pass to the driver.
  --driver-class-path         Extra class path entries to pass to the driver. Note that
                              jars added with --jars are automatically included in the
                              classpath.

  --executor-memory MEM       Memory per executor (e.g. 1000M, 2G) (Default: 1G).

  --proxy-user NAME           User to impersonate when submitting the application.
```

图 11-18　spark-shell 启动及参数一

```
  --proxy-user NAME           User to impersonate when submitting the application.
  --help, -h                  Show this help message and exit
  --verbose, -v               Print additional debug output
  --version,                  Print the version of current Spark

Spark standalone with cluster deploy mode only:
  --driver-cores NUM          Cores for driver (Default: 1).

Spark standalone or Mesos with cluster deploy mode only:
  --supervise                 If given, restarts the driver on failure.
  --kill SUBMISSION_ID        If given, kills the driver specified.
  --status SUBMISSION_ID      If given, requests the status of the driver specified.

Spark standalone and Mesos only:
  --total-executor-cores NUM  Total cores for all executors.

Spark standalone and YARN only:
  --executor-cores NUM        Number of cores per executor. (Default: 1 in YARN mode,
                              or all available cores on the worker in standalone mode)

YARN-only:
  --driver-cores NUM          Number of cores used by the driver, only in cluster mode
                              (Default: 1).
  --queue QUEUE_NAME          The YARN queue to submit to (Default: "default").
  --num-executors NUM         Number of executors to launch (Default: 2).
  --archives ARCHIVES         Comma separated list of archives to be extracted into the
                              working directory of each executor.
  --principal PRINCIPAL       Principal to be used to login to KDC, while running on
                              secure HDFS.
  --keytab KEYTAB             The full path to the file that contains the keytab for the
                              principal specified above. This keytab will be copied to
                              the node running the Application Master via the Secure
                              Distributed Cache, for renewing the login tickets and the
                              delegation tokens periodically.
```

图 11-19　spark-shell 启动及参数二

spark-shell 启动后客户端界面如图 11-20 所示，spark-shell 启动后 YARN 界面图 11-21 所示，spark-shell 启动后 Spark 界面如图 11-22 所示。

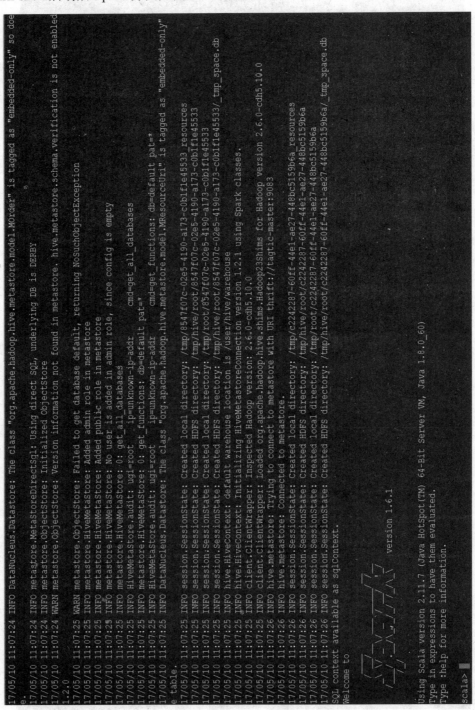

图 11-20　spark-shell 启动后客户端界面

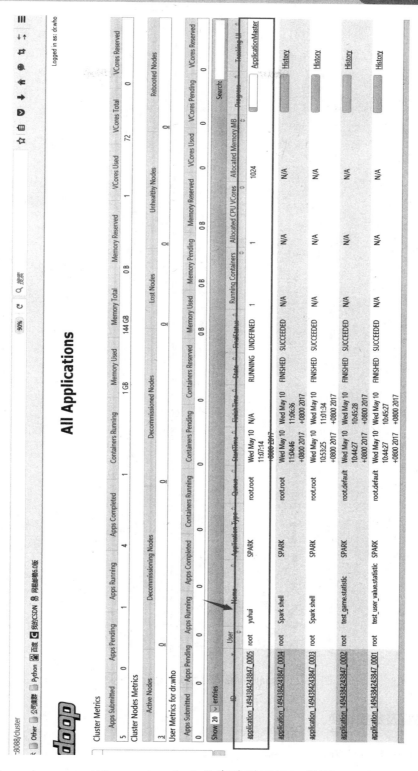

图 11-21 spark-shell 启动后 YARN 界面

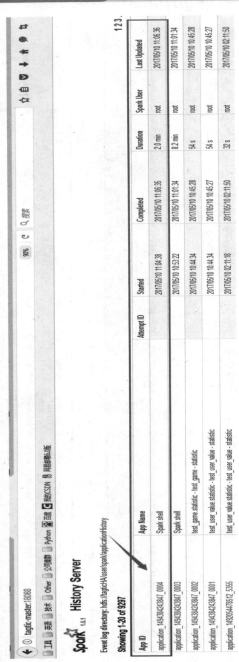

图 11-22 spark-shell 启动后 Spark 界面

4. 加载数据及数据转换

（1）加载 Json 数据转为 DataFrame

```
scala>import sqlContext.implicits._
import sqlContext.implicits.
```

```
scala>val df =
sqlContext.read.json("/user/yuhui/sparksql/people.json")

// Displays the content of the DataFrame to stdout
scala>df.show()
```

结果如图 11-23 所示。

图 11-23

（2）加载文本数据转为 DataFrame

```
spark-shell --driver-memory 1G --executor-memory 1G --executor-cores
4 --name yuhui

// sc is an existing SparkContext.
val sqlContext = new org.apache.spark.sql.SQLContext(sc)
// this is used to implicitly convert an RDD to a DataFrame.
import sqlContext.implicits._

// Define the schema using a case class.
// Note: Case classes in Scala 2.10 can support only up to 22 fields.
To work around this limit,
// you can use custom classes that implement the Product interface.
case class Person(name: String, age: Int)

// Create an RDD of Person objects and register it as a table.
val people =
sc.textFile("/user/yuhui/sparksql/people.txt").map(_.split(",")).map(
p => Person(p(0), p(1).trim.toInt)).toDF()
people.registerTempTable("people")

// SQL statements can be run by using the sql methods provided by
sqlContext.
val teenagers = sqlContext.sql("SELECT name, age FROM people WHERE
age >= 13 AND age <= 19")

// The results of SQL queries are DataFrames and support all the
normal RDD operations.
// The columns of a row in the result can be accessed by field index:
teenagers.map(t => "Name: " + t(0)).collect().foreach(println)
```

结果如图 11-24 所示。

```
Name: Justin
```

图 11-24

```
// or by field name:
teenagers.map(t => "Name: " +
t.getAs[String]("name")).collect().foreach(println)
```

结果如图 11-25 所示。

```
Name: Justin
```

图 11-25

```
// row.getValuesMap[T] retrieves multiple columns at once into a
Map[String, T]
teenagers.map(_.getValuesMap[Any](List("name",
"age"))).collect().foreach(println)
// Map("name" -> "Justin", "age" -> 19)
```

结果如图 11-26 所示。

```
Map(name -> Justin, age -> 19)
```

图 11-26

（3）加载 parquet 转为 DataFrame

```
val df =
sqlContext.read.load("/user/yuhui/sparksql/namesAndAges.parquet")
df.show()
```

结果如图 11-27 所示。

图 11-27

（4）数据转为 parquet 格式

```
val df = sqlContext.read.format("json").load("/user/yuhui/sparksql/
people.json")
df.select("name",
"age").write.format("parquet").save("/user/yuhui/sparksql/namesAndAge
s.
parquet")
```

5. DataFrame 操作

```
// Create the DataFrame
val df = sqlContext.read.json("/user/yuhui/sparksql/people.json")

// Show the content of the DataFrame
df.show()
```

结果如图 11-28 所示。

图 11-28

```
// Print the schema in a tree format
df.printSchema()
```

结果如图 11-29 所示。

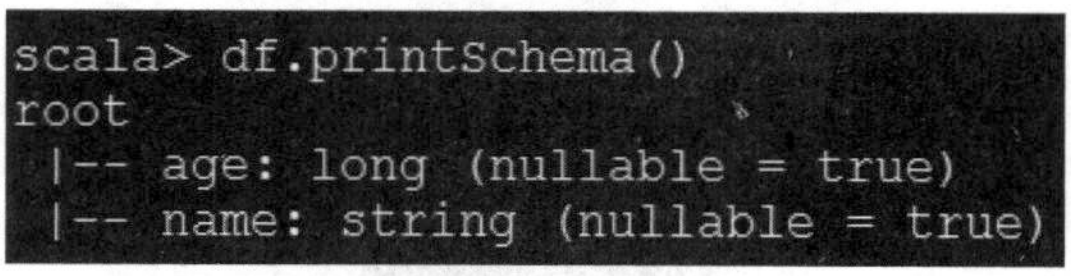

图 11-29

```
// Select only the "name" column
df.select("name").show()
```

结果如图 11-30 所示。

图 11-30

```
// Select everybody, but increment the age by 1
df.select(df("name"), df("age") + 1).show()
```

结果如图 11-31 所示。

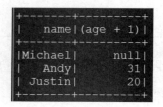

图 11-31

```
// Select people older than 21
df.filter(df("age") > 21).show()
```

结果如图 11-32 所示。

图 11-32

```
// Count people by age
df.groupBy("age").count().show()
```

结果如图 11-33 所示。

图 11-33

6. SparkSql 操作

```
val df = sqlContext.read.json("/user/yuhui/sparksql/people.json")

df.registerTempTable("people")

sqlContext.sql("SELECT name FROM people WHERE age >= 13 AND age <= 19").show()
```

结果如图 11-34 所示。

图 11-34

```
sqlContext.sql("SELECT count(1) FROM people").show()
```

结果如图 11-35 所示。

图 11-35

```
sqlContext.sql("SELECT * FROM people").show()
```

结果如图 11-36 所示。

图 11-36

7. Schema 操作

（1）Schema 合并

```
// sqlContext from the previous example is used in this example.
// This is used to implicitly convert an RDD to a DataFrame.
import sqlContext.implicits._

// Create a simple DataFrame, stored into a partition directory
val df1 = sc.makeRDD(1 to 5).map(i => (i, i * 2)).toDF("single", "double")
df1.write.parquet("/user/yuhui/sparksql/data/test_table/key=1")

// Create another DataFrame in a new partition directory,
// adding a new column and dropping an existing column
val df2 = sc.makeRDD(6 to 10).map(i => (i, i * 3)).toDF("single", "triple")
df2.write.parquet("/user/yuhui/sparksql/data/test_table/key=2")

// Read the partitioned table
val df3 = sqlContext.read.option("mergeSchema", "true").parquet("/user/yuhui/sparksql/data/test_table")
```

结果如图 11-37、图 11-38 所示。

Permission	Owner	Group	Size	Last Modified	Replication	Block Size	Name
/user/yuhui/sparksql/data/test_table/key=1							
-rw-r--r--	root	supergroup	0 B	Wed May 10 15:37:49 +0800 2017	2	128 MB	_SUCCESS
-rw-r--r--	root	supergroup	297 B	Wed May 10 15:37:49 +0800 2017	2	128 MB	_common_metadata
-rw-r--r--	root	supergroup	757 B	Wed May 10 15:37:49 +0800 2017	2	128 MB	_metadata
-rw-r--r--	root	supergroup	531 B	Wed May 10 15:37:47 +0800 2017	2	128 MB	part-r-00000-c5f38ddc-50d1-4f85-bc24-ba759e1253e1.gz.parquet
-rw-r--r--	root	supergroup	535 B	Wed May 10 15:37:49 +0800 2017	2	128 MB	part-r-00001-c5f38ddc-50d1-4f85-bc24-ba759e1253e1.gz.parquet

图 11-37 Schema 合并结果一

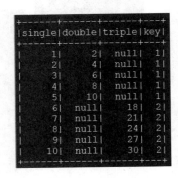

图 11-38 Schema 合并结果二

（2）编程方式指定 Scahema

首先打开一下 people.json、people.txt 两个文件的内容，如表 11-2 所示。

表 11-2 数据文件内容

people.json	people.txt
{"name":"Michael"} {"name":"Andy", "age":30} {"name":"Justin", "age":19}	Michael, 29 Andy, 30 Justin, 19

代码如下：

```
// Create an RDD of Person objects and register it as a table.
val people = sc.textFile("/user/yuhui/sparksql/people.txt")

// The schema is encoded in a string
val schemaString = "name age"

// Import Row.
import org.apache.spark.sql.Row;

// Import Spark SQL data types
import
```

```
org.apache.spark.sql.types.{StructType,StructField,StringType};

// Generate the schema based on the string of schema
val schema = StructType(   schemaString.split(" ").map(fieldName =>
StructField(fieldName, StringType, true)))

// Convert records of the RDD (people) to Rows.
val rowRDD = people.map(_.split(",")).map(p => Row(p(0), p(1).trim))

// Apply the schema to the RDD.
val peopleDataFrame = sqlContext.createDataFrame(rowRDD, schema)

// Register the DataFrames as a table.
peopleDataFrame.registerTempTable("people")

// SQL statements can be run by using the sql methods provided by
sqlContext.
val results = sqlContext.sql("SELECT name FROM people")

// The results of SQL queries are DataFrames and support all the
normal RDD operations.
// The columns of a row in the result can be accessed by field index
or by field name.
results.map(t => "Name: " + t(0)).collect().foreach(println)
```

结果如图 11-39 所示。

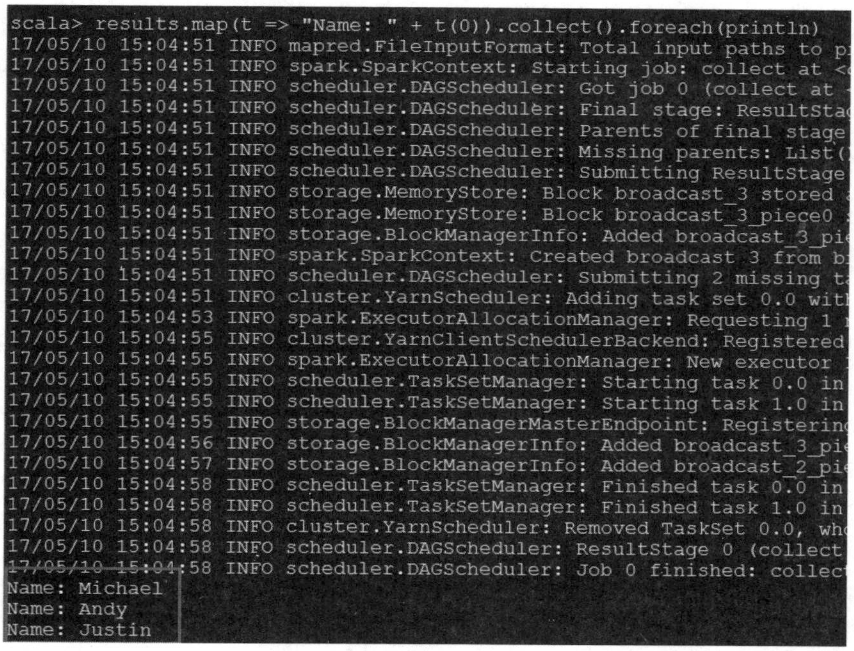

图 11-39　指定 Scahema 结果

11.6 小结

　　Spark 是专为大规模数据处理而设计的快速通用的计算引擎。本章讲解 Spark 概述和基本概念、Spark 的算子、Spark Streaming、Spark SQL、DataFrame。在 Spark 概述和基本概念中介绍了 Spark 的由来、Spark 的生态体系、Spark 与 MapReduce 的关系、Spark 的运行模式、RDD 简介、Spark 的提交方式。Spark 算子中介绍了 Value 型、Key-Value 型、Actions 型三类算子的使用说明及运行结果。Spark Streaming 中介绍了 Spark Streaming 简介、DStreams、Dstreams 的时间窗口、Dstream 的原理和操作。Spark SQL 和 DataFrame 介绍了 3 种数据的加载方式，数据加载完毕之后通过 Spark SQL 和 DataFrame 进行实例操作，之后讲解了 Schema 的操作。

第 12 章 大数据网站日志分析项目

12.1 项目介绍

1. 网站统计产品概述

网站统计，可以实时对网站流量进行分析，帮助网站管理员、运营人员、推广人员等实时获取网站流量，并从流量来源、网站内容、网站访客特性等多方面提供网站分析的数据依据。从而帮助提高网站流量，提升网站用户体验，让访客更多地沉淀下来变成会员或客户，通过更少的投入获取最大化的收入。

2. 网站统计产品的优势

（1）历史数据永久保留

网站统计会一直保留网站的历史统计数据，不进行删除。某些同类产品会进行定期的数据删除，对你的历史数据分析对比会产生极大影响。

（2）7 日完整明细

提供最近 7 天的完整网站浏览明细，并可以通过各种条件复合查询进行筛选、搜索。在产生数据疑问的时候可以通过明细功能进行及时的分析。这在同类产品中是绝无仅有的。

（3）服务专业、细致

网站统计客户服务体系健全，当你对功能的使用、报表的解读有疑问，或者希望提供更符合你需求的功能时，都可以方便地找到。

12.2 网站离线项目

12.2.1 业务框架图

离线项目业务框架如图 12-1 所示。

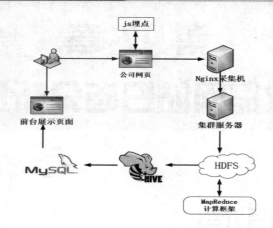

图 12-1　离线项目业务框架

12.2.2　子服务"趋势分析"详解

1. 页面展示

趋势分析页面如图 12-2 所示。

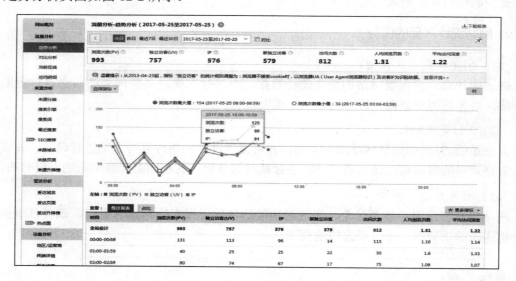

图 12-2　趋势分析页面

2. 功能说明

根据你选定的时间段提供网站流量数据，通过流量的趋势变化形态，可帮助你分析出网站访客的访问规律、网站发展状况等。

（1）了解网站质量和运营状况

如果你网站整体流量偏低，说明你的网站内容不足以吸引访客，或者网站运营不够好。

如果你网站平均访问时长低、跳出率高，说明访客在你的网站找不到感兴趣的内容，或者不会使用，建议你优化网站内容和结构，提高网站质量。

（2）掌握流量规律，制定运营策略

选择按"小时"查看，了解网站一天24小时流量规律。例如你发现某时段流量较高，但不在推广范围内，那么你在进行网站推广时，可以优先考虑该时段，以获得更多潜在用户。每小时网页展示如图12-3所示。

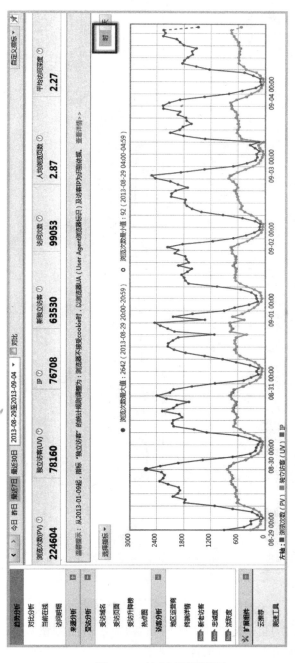

图12-3 每小时网页展示

选择按"天"查看，了解网站一周内各天的流量规律。例如你发现周末网站流量明显高于工作日，那么你在周末进行网站宣传，容易获得更好的推广效果。每天网页展示如图 12-4 所示。

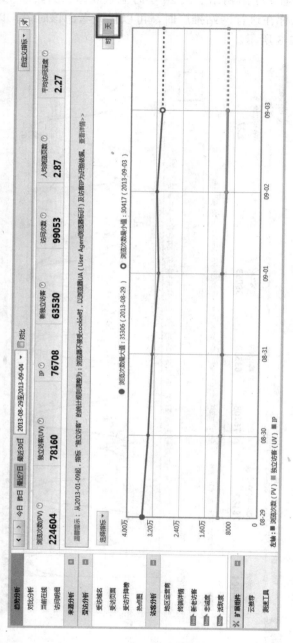

图 12-4　每天网页展示

选择按"周"/按"月"查看，了解网站各周/各月的流量规律。例如通过按"月"查看，你可以快速了解网站不同季度的流量规律，进而针对不同季度开展不同的推广活动。每月网页展示如图 12-5 所示。

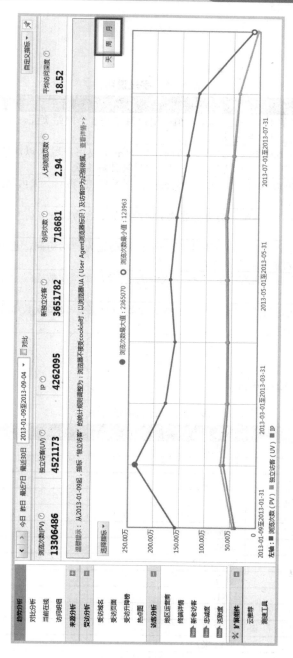

图 12-5　每月网页展示

（3）监控流量起伏，了解运营效果

例如你在某时段开展了网站推广活动，通过趋势分析数据，你可快速直观地了解该活动宣传效果。

（4）及时发现异常，避免流量继续下跌

例如你在趋势分析中发现当前数据异常下跌，于是可以立即排查流量下跌原因，避免网站利益继续受损。

3. 字段说明

字段说明如表 12-1 所示。

表 12-1　字段说明

字段	说明
浏览次数 PV	即通常说的 PV（PageView）值，用户每打开 1 个网站页面，记录 1 个 PV。用户多次打开同一页面 PV 累计多次
独立访客 UV	1 天（00:00-24:00）之内，访问网站的不重复用户数（以浏览器 user_tracecode 为依据），一天内同一访客多次访问网站只被计算 1 次
IP	1 天（00:00-24:00）之内，访问网站的不重复 IP 数。一天内相同 IP 地址多次访问网站只被计算 1 次
新独立访客 Newuser	当日的独立访客中，历史上首次访问网站的访客为新独立访客
访问次数 visittimes	访客从进入网站到离开网站的一系列活动记为一次访问，也称会话(session),1 次访问（会话）可能包含多个 PV
人均浏览页数 Avgpv	平均每个独立访客产生的 PV。人均浏览页数=浏览次数/独立访客。体现网站对访客的吸引程度
平均访问深度 Avgvisittimes	平均每次访问（会话）产生的 PV。平均访问深度=浏览次数/访问次数。体现网站对访客的吸引程度

4. 采集日志说明

采集日志说明如表 12-2 所示。

表 12-2　日志字段说明

序号	日志字段	说明
0	timestamp	时间蹉
1	user_ip	IP
2	user_tracecode	用户 id
3	user_sessionid	会话
4	domain	域名
5	screen_width	屏幕宽
6	screen_height	屏幕高
7	color_dept	颜色深度
8	language	语言
9	url	URL
10	referrer	来源
11	user_agent	浏览器
12	account	追踪账户
13	event	事件（单击，鼠标移动，打开）
14	event_data	事件数据

12.2.3　表格的设计

1. Hive 表格字段设计

Hive 表格字段描述如表 12-3 所示。

表 12-3 Hive 表格字段说明

字段	字段名称	字段类型
timestamp	时间蹉	Varchar
user_ip	IP	Varchar
user_tracecode	用户 id	Varchar
user_sessionid	会话	Varchar
domain	域名	Varchar
screen_width	屏幕宽	Varchar
screen_height	屏幕高	Varchar
color_dept	颜色深度	Varchar
language	语言	Varchar
url	URL	Varchar
referrer	来源	Varchar
user_agent	浏览器	Varchar
account	追踪账户	Varchar
event	事件（单击，鼠标移动，打开）	Varchar
event_data	事件数据	Varchar
day	天	Varchar
hour	小时	Varchar

2. Hive 建立表格

```
hive> CREATE EXTERNAL TABLE logtable(timestamp string,user_ip
string,user_tracecode string,user_sessionid string,domain
string,screen_width string,screen_height string,color_dept
string,language string,url string,referrer string,user_agent
string,account string,event string,event_data string)partitioned by (day
string,hour string) row format delimited fields terminated by '\t';
```

3. MySQL 表格字段设计

MySQL 表格字段设计如表 12-4 所示。

表 12-4 MySQL 表格字段

字段	字段名称	字段类型
Day	天	String
Hour	小时	String
PV	浏览次数(PV)	Int
UV	独立访客(UV)	Int
IP	IP 数	Int
Newuser	新独立访客	Int
visittimes	访问次数	Int
Avgpv	人均浏览页数	Double
Avgvisittimes	平均访问深度	Double

4. 建立 MySQL 表格

```
CREATE TABLE logtable(
    id int not null auto_increment,
    day VARCHAR(45) null,
```

```
    hour VARCHAR(45) null,
    pv INT NULL,
    uv INT NULL,
    ip INT NULL,
    newuser INT NULL,
    visittimes INT NULL,
    avgpv Double NULL,
    avgvisittimes Double NULL,
    primary key (id ,day ,hour)
)ENGINE=InnoDB DEFAULT CHARSET=utf8;
```

12.2.4 提前准备

网站离线项目的编辑器是 Eclipse，编程语言是 Java，环境是 CDH5.10.0。在指标计算中只计算 Day、Hour、PV、UV、IP，其余指标计算读者可以自己操作，本机主要以数据走向和框架介绍为主。

1. 下载客户端配置

提前在 CDH 中下载客户端配置，放在 resources 中，程序启动时候，需要预加载 resources 中的文件到 Configuration 类中。客户端配置下载如图 12-6 所示。

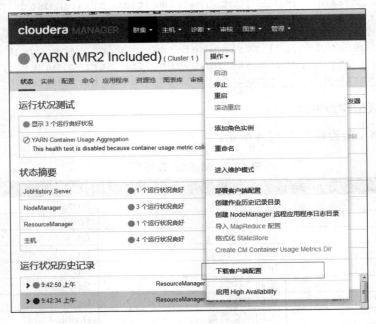

图 12-6　客户端配置下载

2. 项目代码展示

项目代码展示如图 12-7 所示。

```
▲ 🏛 Maven_BigData
    ▲ 🗁 src/main/java
        ▷ 🌐 cn.orcale.com.bigdata.hbase
        ▷ 🌐 cn.orcale.com.bigdata.hdfs
        ▷ 🌐 cn.orcale.com.bigdata.hive
        ▷ 🌐 cn.orcale.com.bigdata.mr.basicmr
        ▷ 🌐 cn.orcale.com.bigdata.mr.customgroup
        ▷ 🌐 cn.orcale.com.bigdata.mr.groupdefault
        ▷ 🌐 cn.orcale.com.bigdata.mr.partition01
        ▷ 🌐 cn.orcale.com.bigdata.mr.partitionsolr
        ▷ 🌐 cn.orcale.com.bigdata.mr.secondsolr
        ▷ 🌐 cn.orcale.com.bigdata.mr.solrkey
        ▷ 🌐 cn.orcale.com.bigdata.test
        ▷ 🌐 cn.orcale.com.bigdata.utils
        ▷ 🌐 cn.orcale.com.bigdata.zk
        ▲ 🌐 cn.orcale.com.project
            ▷ 🗋 CDHHiveJdbcCli.java
            ▷ 🗋 Logs.java
            ▷ 🗋 mrClean.java
            ▷ 🗋 mysqlCli2.java
    ▲ 🗁 src/main/resources
        🗋 core-site.xml
        🗋 hdfs-site.xml
        🗋 hive-site.xml
        🗋 mapred-site.xml
        🗋 yarn-site.xml
    ▷ 🗁 src/test/java
    ▷ 🗁 src/test/resources
    ▷ 📚 JRE System Library [jdk1.8.0_77]
    ▷ 📚 Maven Dependencies
    ▷ 📦 hive-exec-0.12.0.jar
    ▷ 📦 hive-jdbc-0.12.0.jar
    ▷ 📦 hive-metastore-0.12.0.jar
    ▷ 📦 hive-service-0.12.0.jar
    ▷ 📦 commons-logging-1.1.1.jar
    ▷ 📦 httpclient-4.2.5.jar
    ▷ 📦 httpcore-4.2.4.jar
    ▷ 📦 libfb303-0.9.0.jar
    ▷ 📦 log4j-1.2.16.jar
    ▷ 📦 slf4j-api-1.6.1.jar
    ▷ 📦 mysql-connector-java-5.1.36-bin.jar
```

图 12-7 项目代码展示

3. Eclipse 的 POM 文件

```
<project xmlns="http://maven.apache.org/POM/4.0.0"
xmlns:xsi="http://www.w3.org/2001/XMLSchema-instance"
xsi:schemaLocation="http://maven.apache.org/POM/4.0.0
```

```xml
http://maven.apache.org/xsd/maven-4.0.0.xsd">
  <modelVersion>4.0.0</modelVersion>
  <groupId>cn.oracle.com</groupId>
  <artifactId>Maven_BigData</artifactId>
  <version>0.0.1-SNAPSHOT</version>
  <name>Maven_BigData</name>

    <repositories>
   <!-- cloudera 资源库 -->
       <repository>
       <id>cloudera</id>
       <url>https://repository.cloudera.com/artifactory/cloudera-repos/</url>
       </repository>
    </repositories>

<dependencies>

        <!-- 测试单元 -->
        <dependency>
            <groupId>junit</groupId>
            <artifactId>junit</artifactId>
            <version>4.12</version>
        </dependency>

        <!-- Hadoop 中公用组件 -->
        <dependency>
            <groupId>org.apache.hadoop</groupId>
            <artifactId>hadoop-common</artifactId>
            <version>2.6.0-cdh5.4.2</version>
        </dependency>

        <!-- HDFS -->
        <dependency>
            <groupId>org.apache.hadoop</groupId>
            <artifactId>hadoop-hdfs</artifactId>
            <version>2.6.0-cdh5.4.2</version>
        </dependency>


        <!-- MapReduce -->
        <dependency>
            <groupId>org.apache.hadoop</groupId>
            <artifactId>hadoop-mapreduce-client-app</artifactId>
            <version>2.6.0-cdh5.4.2</version>
        </dependency>
        <dependency>
            <groupId>org.apache.hadoop</groupId>
            <artifactId>hadoop-mapreduce-client-common</artifactId>
            <version>2.6.0-cdh5.4.2</version>
```

```xml
        </dependency>
        <dependency>
            <groupId>org.apache.hadoop</groupId>
            <artifactId>hadoop-mapreduce-client-core</artifactId>
            <version>2.6.0-cdh5.4.2</version>
        </dependency>
         <dependency>
            <groupId>org.apache.hadoop</groupId>
            <artifactId>hadoop-mapreduce-client-jobclient</artifactId>
            <version>2.6.0-cdh5.4.2</version>
        </dependency>

        <!-- YARN -->
        <dependency>
            <groupId>org.apache.hadoop</groupId>
            <artifactId>hadoop-yarn-api</artifactId>
            <version>2.6.0-cdh5.4.2</version>
         </dependency>
        <dependency>
            <groupId>org.apache.hadoop</groupId>
            <artifactId>hadoop-yarn-client</artifactId>
            <version>2.6.0-cdh5.4.2</version>
        </dependency>
        <dependency>
            <groupId>org.apache.hadoop</groupId>
            <artifactId>hadoop-yarn-common</artifactId>
            <version>2.6.0-cdh5.4.2</version>
        </dependency>
        <dependency>
            <groupId>org.apache.hadoop</groupId>
            <artifactId>hadoop-yarn-applications-distributedshell</artifactId>
            <version>2.6.0-cdh5.4.2</version>
        </dependency>


        <!-- Hbase 的 api -->
        <dependency>
            <groupId>org.apache.hbase</groupId>
            <artifactId>hbase-client</artifactId>
            <version>1.0.0-cdh5.4.2</version>
        </dependency>

        <dependency>
            <groupId>org.apache.hbase</groupId>
            <artifactId>hbase-common</artifactId>
            <version>1.0.0-cdh5.4.2</version>
        </dependency>
```

```xml
        <dependency>
            <groupId>org.apache.hbase</groupId>
            <artifactId>hbase-testing-util</artifactId>
            <version>1.0.0-cdh5.4.2</version>
        </dependency>

        <dependency>
            <groupId>org.apache.hbase</groupId>
            <artifactId>hbase-shell</artifactId>
            <version>1.0.0-cdh5.4.2</version>
        </dependency>

    <!-- 解决 Maven 工程中报 Missing artifact jdk.tools:jdk.tools: -->
        <dependency>
            <groupId>jdk.tools</groupId>
            <artifactId>jdk.tools</artifactId>
            <version>1.6</version>
            <scope>system</scope>
            <systemPath>${JAVA_HOME}/lib/tools.jar</systemPath>
        </dependency>



</dependencies>

</project>
```

4. 建立日志清洗类

```java
package cn.orcale.com.project;

import java.io.IOException;

import org.apache.commons.lang.StringUtils;
import org.apache.hadoop.conf.Configuration;
import org.apache.hadoop.fs.FileSystem;
import org.apache.hadoop.fs.Path;
import org.apache.hadoop.io.Text;
import org.apache.hadoop.mapreduce.Job;
import org.apache.hadoop.mapreduce.Mapper;
import org.apache.hadoop.mapreduce.Reducer;
import org.apache.hadoop.mapreduce.lib.input.FileInputFormat;
import org.apache.hadoop.mapreduce.lib.output.FileOutputFormat;

import cn.orcale.com.bigdata.utils.DateUtils;

//cn.orcale.com.project.mrClean
public class mrClean {


    //map 将输入中的 value 复制到输出数据的 key 上，并直接输出
```

```java
    public static class Map extends Mapper<Object,Text,Text,Text>{
        //实现 map 函数
        public void map(Object key,Text value,Context context) throws IOException,InterruptedException{
            DateUtils du = new DateUtils();
            String day = "";
            String hour = "";
            String[] fields = StringUtils.split(value.toString() , "");
            StringBuffer sb = new StringBuffer();
            // 过滤字段少的数据
            if (fields.length < 15) {
                return;
            }

            // 过滤 domain 为空
            if (StringUtils.isBlank(fields[4])) {
                return;
            }

            day = du.stampToDate(fields[0].substring(0,10)+"000").substring(0,10);

            hour = du.stampToDate(fields[0].substring(0,10)+"000").substring(11,13);

            for(int i = 0; i<fields.length ; i++){
                sb.append(fields[i]+'\t');
            }

            sb.append(day +'\t' + hour) ;
        context.write(new Text(sb.toString()), new Text(""));
        }
    }


    //reduce 将输入中的 key 复制到输出数据的 key 上,并直接输出
    public static class Reduce extends Reducer<Text,Text,Text,Text>{
        //实现 reduce 函数
        public void reduce(Text key,Iterable<Text> values,Context
```

```java
context)
            throws IOException,InterruptedException{
        context.write(key, new Text(""));
    }
}


    public static void main(String[] args) throws Exception{
    Configuration conf = new Configuration();
        Job job = Job.getInstance(conf);
        job.setJarByClass(mrClean.class);
        //设置Map、Combine 和 Reduce 处理类
        job.setMapperClass(Map.class);
        job.setReducerClass(Reduce.class);
        //指定 reduce task 数量，跟 ProvincePartitioner 的分区数匹配
        job.setNumReduceTasks(1);

        //本次 job 作业 mapper 类的输出数据 key 类型
        job.setMapOutputKeyClass(Text.class);
        //本次 job 作业 mapper 类的输出数据 value 类型
        job.setMapOutputValueClass(Text.class);

        //本次 job 作业 reducer 类的输出数据 key 类型
        job.setOutputKeyClass(Text.class);
        //本次 job 作业 reducer 类的输出数据 value 类型
        job.setOutputValueClass(Text.class);

        String inputPath = "/user/yuhui/logs/ma-"+args[0]+".log";
        String ddate=args[0].substring(0, 8);//2017052610
        System.out.println("ddate:="+ddate);
        String hour=args[0].substring(8,args[0].length());
        System.out.println("hour:="+hour);
        String outputPath = "/user/yuhui/cleanlogs/"+ddate+"/"+hour;


        // 判断 output 文件夹是否存在，如果存在则删除
        Path path = new Path(outputPath);// 取第1个表示输出目录参数（第0个参数是输入目录）
        FileSystem fileSystem = path.getFileSystem(conf);// 根据 path 找到这个文件
        if (fileSystem.exists(path)) {
            fileSystem.delete(path, true);// true 的意思是，就算 output 有东西，也一带删除
        }
```

```
//        //本次job作业要处理的原始数据所在的路径
        FileInputFormat.setInputPaths(job, new Path(inputPath));
//        //本次job作业产生的结果输出路径
        FileOutputFormat.setOutputPath(job, new Path(outputPath));
        job.waitForCompletion(true);
    }
}
```

5. 建立 Logs 对象

```java
package cn.orcale.com.project;

public class Logs {

    String DAY ;

    String HOUR;

    int PV;

    int UV;

    int IP;

    int Newuser;

    int VisitTimes;

    Double Avgpv;

    Double Avgvisittimes;

    public Logs() {
        super();
        // TODO Auto-generated constructor stub
    }

    public Logs(String dAY, String hOUR, int pV, int uV, int iP, int newuser,
            int visitTimes, Double avgpv, Double avgvisittimes) {
        super();
        DAY = dAY;
        HOUR = hOUR;
        PV = pV;
        UV = uV;
        IP = iP;
        Newuser = newuser;
        VisitTimes = visitTimes;
        Avgpv = avgpv;
        Avgvisittimes = avgvisittimes;
    }
```

```java
    public String getDAY() {
        return DAY;
    }

    public void setDAY(String dAY) {
        DAY = dAY;
    }

    public String getHOUR() {
        return HOUR;
    }

    public void setHOUR(String hOUR) {
        HOUR = hOUR;
    }

    public int getPV() {
        return PV;
    }

    public void setPV(int pV) {
        PV = pV;
    }

    public int getUV() {
        return UV;
    }

    public void setUV(int uV) {
        UV = uV;
    }

    public int getIP() {
        return IP;
    }

    public void setIP(int iP) {
        IP = iP;
    }

    public int getNewuser() {
        return Newuser;
    }

    public void setNewuser(int newuser) {
        Newuser = newuser;
    }

    public int getVisitTimes() {
```

```
        return VisitTimes;
    }

    public void setVisitTimes(int visitTimes) {
        VisitTimes = visitTimes;
    }

    public Double getAvgpv() {
        return Avgpv;
    }

    public void setAvgpv(Double avgpv) {
        Avgpv = avgpv;
    }

    public Double getAvgvisittimes() {
        return Avgvisittimes;
    }

    public void setAvgvisittimes(Double avgvisittimes) {
        Avgvisittimes = avgvisittimes;
    }


}
```

6. 建立 Java 的 Hive 连接

```
package cn.orcale.com.project;

import java.sql.Connection;
import java.sql.DriverManager;
import java.sql.ResultSet;
import java.sql.SQLException;
import java.sql.Statement;

/**
 * Hive 的 JavaApi
 *
 */
public class CDHHiveJdbcCli {

    private static String driverName = "org.apache.hive.jdbc.HiveDriver";
    private static String url = "jdbc:hive2://tagtic-master:10000/default";
    private static String user = "root";
    private static String password = "tagtic-master";

    public static void main (String[] args) {
```

```java
        try {
                Class.forName(driverName);
                Connection conn = DriverManager.getConnection(url,user,password);
                Statement stmt = conn.createStatement();

                String sql=" select 20170526 ,19 , count(1) as pv , size(collect_set(user_tracecode ) ) as uv , "
                        + "size(collect_set(user_ip ) ) as IP , 0 ,0 , 0 ,0 from logtable "
                        + "where day='20170526' and hour='19'";

                ResultSet res = stmt.executeQuery(sql);

                System.out.println("DAY"+ "\t"+ "hour"+ "\t" + "PV"+ "\t" + "UV" + "\t"+ "IP" + "\t"+ "Newuser"
                        + "\t"+ "VisitTimes" + "\t"+ "Avgpv" + "\t"+ "Avgvisittimes");
                Logs logs = new Logs();
                while (res.next()) {
                    System.out.println(res.getString(1) + "\t" + res.getString(2)+ "\t" +
                                        res.getString(3)+ "\t" + res.getString(4)+ "\t" +
                                        res.getString(5)+ "\t" + res.getString(6)+ "\t" +
                                        res.getString(7));

                    logs.DAY=res.getString(1);

                    logs.HOUR=res.getString(2);

                    logs.PV=Integer.parseInt(res.getString(3));

                    logs.UV=Integer.parseInt(res.getString(4));

                    logs.IP=Integer.parseInt(res.getString(5));

                    logs.Newuser=Integer.parseInt(res.getString(6));

                    logs.VisitTimes=Integer.parseInt(res.getString(7));

                    logs.Avgpv=Double.parseDouble(res.getString(8));

            logs.Avgvisittimes=Double.parseDouble(res.getString(9));

                    mysqlCli2.InsertSql(logs);
                }
```

```
                    conn.close();
                    conn = null;
                    
                    System.out.println("Hive 操作完毕");
            } catch (ClassNotFoundException e) {
                e.printStackTrace();
                System.exit(1);
            } catch (SQLException e) {
                e.printStackTrace();
                System.exit(1);
            }
        }
    }
}
```

7. 建立 Java 的 MySQL 连接

```
package cn.orcale.com.project;

import java.sql.*;

public class mysqlCli2 {

    // 驱动程序名
    static String driver = "com.mysql.jdbc.Driver";

    // URL 指向要访问的数据库名 scutcs
    static String url = "jdbc:mysql://tagtic-master:3306/yuhui";

    // MySQL 配置时的用户名
    static String user = "root";

    // MySQL 配置时的密码
    static String password = "tagtic-master";
//    public static void main(String[] args) throws Exception{
//
//        mysqlCli mysqlCli = new mysqlCli();
//        Logs logs = new Logs();
//        logs.DAY="20170527";
//        logs.HOUR="14";
//        logs.PV=100;
//        logs.UV=200;
//        logs.IP=300;
//        logs.Newuser=400;
//        logs.VisitTimes=500;
//        logs.Avgpv=500.0;
```

```java
//         logs.Avgvisittimes=600.0;
//         mysqlCli2.InsertSql(logs);
//
//     }

    public static void InsertSql(Logs log){

    try {

        Connection conn = null ;

            Class.forName(driver);

            conn = DriverManager.getConnection(url, user, password);

            Statement stmt = conn.createStatement();

            String insql="insert into
logtable(day,hour,pv,uv,ip,newuser,visittimes,Avgpv,Avgvisittimes) "
            + "values(?,?,?,?,?,?,?,?,?)";

            PreparedStatement ps=conn.prepareStatement(insql);

            ps.setString(1, log.getDAY());

            ps.setString(2, log.getHOUR());

            ps.setInt(3, log.getPV());

            ps.setInt(4, log.getUV());

            ps.setInt(5, log.getIP());

            ps.setInt(6, log.getNewuser());

            ps.setInt(7, log.getVisitTimes());

            ps.setDouble(8, log.getAvgpv());

            ps.setDouble(9, log.getAvgvisittimes());

            ps.executeUpdate();

            System.out.println("mysql 数据插入完毕");

            stmt.close();
            conn.close();


        } catch (Exception e) {
```

```
                e.printStackTrace();
        }
    }
}
```

8. 使用组件及软件

MapReduce、HDFS、Hive、Eclipse、Java、MySQL。

12.2.5　项目步骤

1. 日志埋点

（1）数据收集原理分析

简单来说，网站统计分析工具需要收集到用户浏览目标网站的行为（如打开某网页、单击某按钮、将商品加入购物车等）及行为附加数据（如某下单行为产生的订单金额等）。早期的网站统计往往只收集一种用户行为：页面的打开，而后用户在页面中的行为均无法收集。这种收集策略能满足基本的流量分析、来源分析、内容分析及访客属性等常用分析视角，但是，随着 Ajax 技术的广泛使用及电子商务网站对于电子商务目标的统计分析的需求越来越强烈，这种传统的收集策略已经显得力不能及。

后来，Google 在其产品谷歌分析中创新性地引入了可定制的数据收集脚本，用户通过谷歌分析定义好的可扩展接口，只需编写少量的 JavaScript 代码就可以实现自定义事件和自定义指标的跟踪和分析。目前百度统计、搜狗分析等产品均照搬了谷歌分析的模式。

其实说起来两种数据收集模式的基本原理和流程是一致的，只是后一种通过 JavaScript 收集到了更多的信息。下面看一下现在各种网站统计工具的数据收集基本原理。

（2）流程概览

首先通过一幅图总体看一下数据收集的基本流程。日志埋点流程如图 12-8 所示。

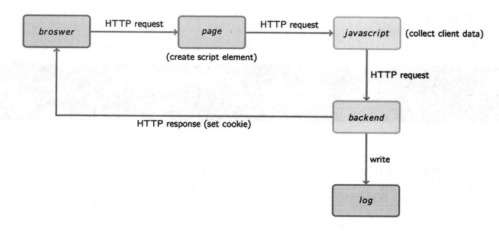

图 12-8　日志埋点流程

首先，用户的行为会触发浏览器对被统计页面的一个 HTTP 请求，这里姑且先认为行为就是打开网页。当网页被打开，页面中的埋点 JavaScript 片段会被执行，用过相关工具的朋友应该知道，一般网站统计工具都会要求用户在网页中加入一小段 JavaScript 代码，这个代码片段一般会动态创建一个 script 标签，并将 src 指向一个单独的 JS 文件，此时这个单独的 JS 文件会被浏览器请求并执行，这个 JS 往往就是真正的数据收集脚本。数据收集完成后，JS 会请求一个后端的数据收集脚本（图 12-8 中的 backend），这个脚本一般是一个伪装成图片的动态脚本程序，可能由 PHP、Python 或其他服务端语言编写，JS 会将收集到的数据通过 HTTP 参数的方式传递给后端脚本，后端脚本解析参数并按固定格式记录到访问日志，同时可能会在 HTTP 响应中给客户端种植一些用于追踪的 Cookie。

2. 埋点代码和字段解析

```
<script type="text/javascript">
var _maq = _maq || [];
_maq.push(['_setAccount', '网站标识']);

(function() {
var ma = document.createElement('script'); ma.type =
'text/javascript'; ma.async = true;
ma.src = ('https:' == document.location.protocol ?
'https://analytics' : 'http://analytics') + '.codinglabs.org/ma.js';
var s = document.getElementsByTagName('script')[0];
s.parentNode.insertBefore(ma, s);
})();
</script>
```

3. 数据采集服务器

Nginx 服务器收集原始数据，每一小时切割一个文件，并且发送到集群服务器。同时每天零点将昨天的日志进行压缩保存，日志名一定要规范按照 YYYY-MM-DD.log 格式，便于以后容错机制的日志恢复。

4. 集群服务器

（1）日志展示（Nginx 每小时日志如图 12-9 所示）

```
[root@tagtic-slave03 projectdata]# ls
ma-2017052613.log  ma-2017052616.log  ma-2017052619.log  ma-2017052622.log
ma-2017052614.log  ma-2017052617.log  ma-2017052620.log  ma-2017052623.log
ma-2017052615.log  ma-2017052618.log  ma-2017052621.log  ma-2017052624.log
```

图 12-9　Nginx 每小时日志

（2）定时上传数据到 HDFS（Nginx 每小时日志存储到 HDFS，如图 12-10 所示）

```
hadoop fs -put ma-20170526* /user/yuhui/logs/
```

Permission	Owner	Group	Size	Last Modified	Replication	Block Size	Name
-rw-r--r--	root	supergroup	1.23 MB	Fri May 26 10:21:57 +0800 2017	2	128 MB	ma-2017052613.log
-rw-r--r--	root	supergroup	2.22 MB	Fri May 26 10:21:58 +0800 2017	2	128 MB	ma-2017052614.log
-rw-r--r--	root	supergroup	1.98 MB	Fri May 26 10:21:58 +0800 2017	2	128 MB	ma-2017052615.log
-rw-r--r--	root	supergroup	1.11 MB	Fri May 26 10:21:58 +0800 2017	2	128 MB	ma-2017052616.log
-rw-r--r--	root	supergroup	1.12 MB	Fri May 26 10:21:58 +0800 2017	2	128 MB	ma-2017052617.log
-rw-r--r--	root	supergroup	428.58 KB	Fri May 26 10:21:58 +0800 2017	2	128 MB	ma-2017052618.log
-rw-r--r--	root	supergroup	1.79 MB	Fri May 26 10:21:58 +0800 2017	2	128 MB	ma-2017052619.log
-rw-r--r--	root	supergroup	1.09 MB	Fri May 26 10:21:58 +0800 2017	2	128 MB	ma-2017052620.log
-rw-r--r--	root	supergroup	4.01 MB	Fri May 26 10:21:58 +0800 2017	2	128 MB	ma-2017052621.log
-rw-r--r--	root	supergroup	2.21 MB	Fri May 26 10:21:58 +0800 2017	2	128 MB	ma-2017052622.log
-rw-r--r--	root	supergroup	153.37 KB	Fri May 26 10:21:58 +0800 2017	2	128 MB	ma-2017052623.log
-rw-r--r--	root	supergroup	4.95 MB	Fri May 26 10:21:59 +0800 2017	2	128 MB	ma-2017052624.log

图 12-10　Nginx 每小时日志存储到 HDFS

5. MR 计算

（1）执行步骤

```
hadoop fs -mkdir /user/yuhui/cleanlogs
hadoop jar LogClean.jar  2017052619
hadoop jar LogClean.jar  2017052620
```

（2）日志清洗之后目录和结果展示

日志清洗之后目录展示如图 12-11~图 12-13 所示，日志清洗之后日志查看如图 12-14 所示。

Browse Directory

/user/yuhui/cleanlogs/20170526

Permission	Owner	Group	Size	Last Modified	Replication	Block Size	Name
drwxr-xr-x	root	supergroup	0 B	Fri May 26 13:47:56 +0800 2017	0	0 B	19
drwxr-xr-x	root	supergroup	0 B	Fri May 26 11:28:11 +0800 2017	0	0 B	20

图 12-11　日志清洗之后目录展示一

Browse Directory

/user/yuhui/cleanlogs/20170526/19

Permission	Owner	Group	Size	Last Modified	Replication	Block Size	Name
-rw-r--r--	root	supergroup	0 B	Fri May 26 13:47:56 +0800 2017	2	128 MB	_SUCCESS
-rw-r--r--	root	supergroup	137.74 KB	Fri May 26 13:47:56 +0800 2017	2	128 MB	part-r-00000

图 12-12　日志清洗之后目录展示二

图 12-13 日志清洗之后目录展示三

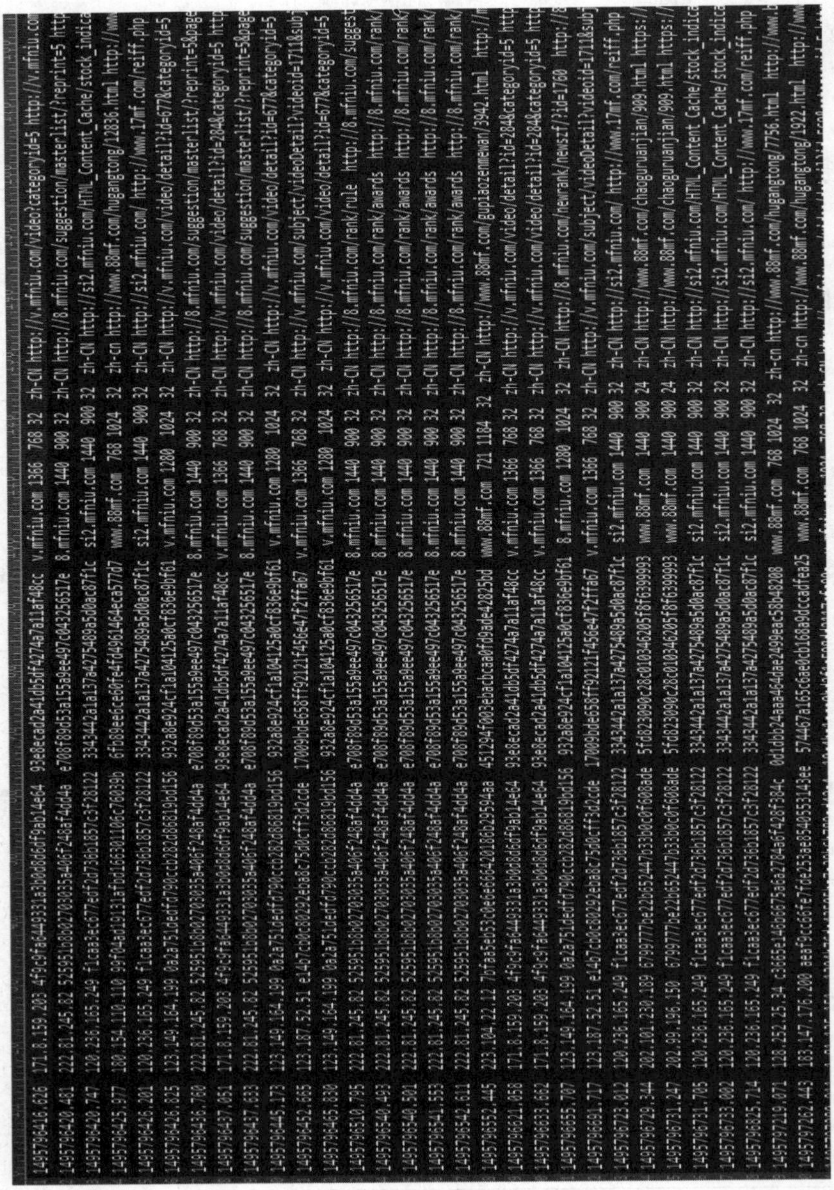

图 12-14 日志清洗之后日志查看

6. Hive 计算，推送到 MySQL

启动 Hive。Hive 客户端启动界面如图 12-15 所示。

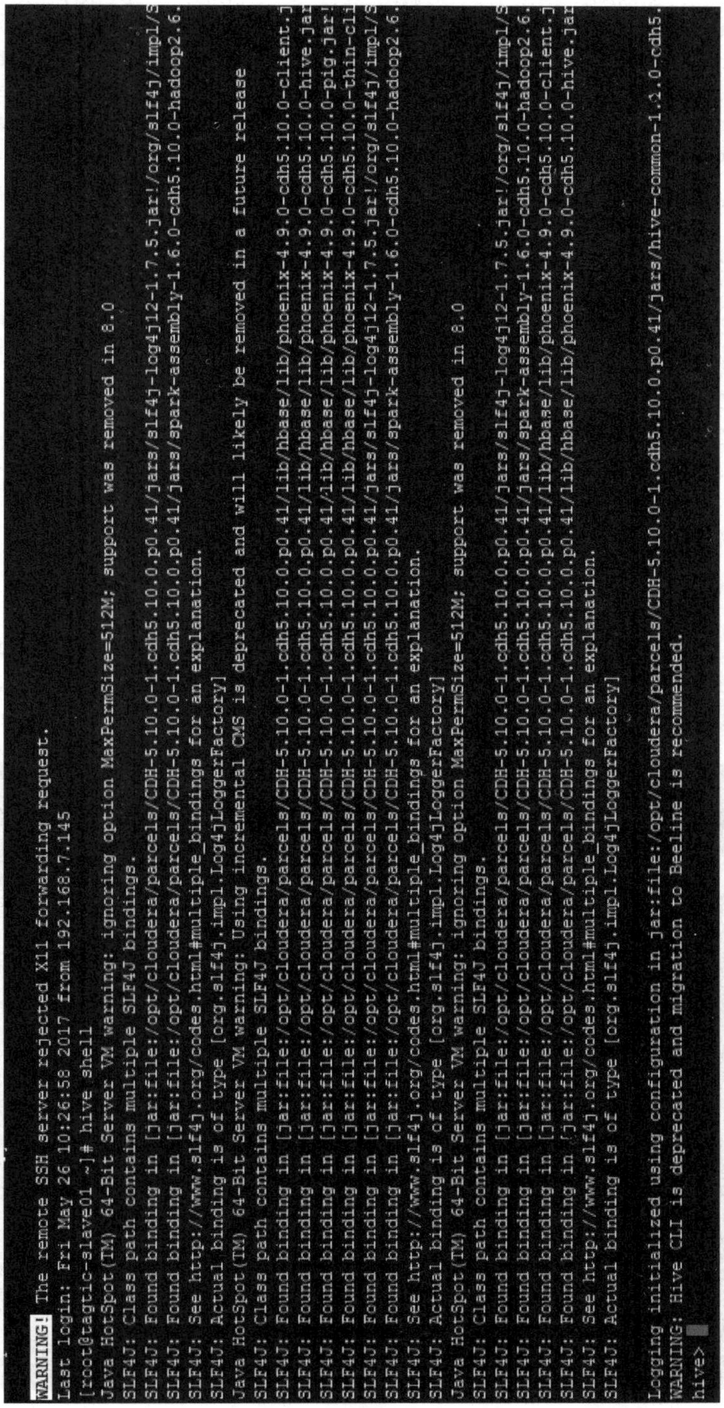

图 12-15　Hive 客户端启动界面

Hive 建立表格，追加数据，查询数据。

```
// Hive 建立表格
hive> CREATE EXTERNAL TABLE logtable(timestamp string,user_ip
string,user_tracecode string,user_sessionid string,domain
string,screen_width string,screen_height string,color_dept
string,language string,url string,referrer string,user_agent
string,account string,event string,event_data string)partitioned by
(day string,hour string) row format delimited fields terminated by
'\t';

//追加数据
hive> ALTER TABLE logtable ADD IF NOT EXISTS
PARTITION(day=20170526,hour=19)LOCATION
'/user/yuhui/cleanlogs/20170526/19;

hive> ALTER TABLE logtable ADD IF NOT EXISTS
PARTITION(day=20170526,hour=20)LOCATION
'/user/yuhui/cleanlogs/20170526/20';

//查询数据
hive> select * from logtable where day='20170526' and hour='19' limit
2;

hive> select * from logtable where day='20170526' and hour='20' limit
2;
```

查询结果如图 12-16 所示。

```
hive> select * from logtable where day='20170526' and hour='19' limit 2;
OK
1495794792.178  125.122.210.196 64d02d6b1530b6ff1172b1ee878e7096        fa11d1b492f1cdf4a
7f9b88dbaa70600 8.wybi.net      1440    900     32      zh-CN   http://8.wybi.net/Promoti
on/videoroom_per_batchstatic.aspx?batch=52      http://8.wybi.net/Promotion/videoroom_per
_batchstatic.aspx?batch=52      Mozilla/5.0 (Windows NT 5.1) AppleWebKit/537.36 (KHTML, l
ike Gecko) Chrome/31.0.1650.63 Safari/537.36    mawenbo _trackClick     648|558|null    2
0170526 19
1495794795.426  125.122.210.196 64d02d6b1530b6ff1172b1ee878e7096        fa11d1b492f1cdf4a
7f9b88dbaa70600 8.wybi.net      1440    900     32      zh-CN   http://8.wybi.net/Promoti
on/videoroom_per_batchstatic.aspx?batch=52      http://8.wybi.net/Promotion/videoroom_per
_batchstatic.aspx?batch=52      Mozilla/5.0 (Windows NT 5.1) AppleWebKit/537.36 (KHTML, l
ike Gecko) Chrome/31.0.1650.63 Safari/537.36    mawenbo _trackMover     723,441|662,457|6
20,499|611,511|588,549|589,572|589,576|636,573|641,569|641,565|661,545|661,538|667,479|67
3,460|681,424|687,387|691,388|691,389|691,389|692,390   20170526        19
Time taken: 0.113 seconds, Fetched: 2 row(s)
hive> select * from logtable where day='20170526' and hour='20' limit 2;
OK
1495796416.626  171.8.159.203   4f9c9fa6449331a3b0d8d6df9ab14e64        93e8ecad2a41db5df
4274a7a11af48cc v.mfniu.com     1366    768     32      zh-CN   http://v.mfniu.com/video?
categoryid=5    http://v.mfniu.com/search?keyword=%E6%90%9E%E7%AC%91    Mozilla/5.0 (Wind
ows NT 6.1) AppleWebKit/537.36 (KHTML, like Gecko) Chrome/30.0.1599.66 Safari/537.36    s
hipindianbo     _trackMover     780,506|786,477|586,491|562,502|485,550|431,592|324,577|3
24,576|322,538|319,526|314,507|312,501|280,477|298,410|308,391|321,368|326,358|329,346|30
8,275|304,261   20170526        20
1495796420.481  222.81.245.82   525951b8b02703035a406f248af4dd4a        e708f89d53a155a9e
e497c043256517e 8.mfniu.com     1440    900     32      zh-CN   http://8.mfniu.com/sugges
tion/masterlist/?reprint=5      http://8.mfniu.com/suggestion/masterdetail/?aid=2577    M
ozilla/5.0 (Windows NT 5.1) AppleWebKit/537.36 (KHTML, like Gecko) Chrome/31.0.1650.48 Sa
fari/537.36 QQBrowser/8.2.4257.400      jiaoyidasai     _trackClick     970|522|http://8.
mfniu.com/suggestion/masterlist/?reprint=5&page=2       20170526        20
Time taken: 0.098 seconds, Fetched: 2 row(s)
hive>
```

图 12-16　查询结果

以计算 PV、UV、IP 为例子，计算方法如下。

```
hive> select 20170526 ,19 , count(1) as pv ,
size(collect_set(user_tracecode ) ) as uv ,
size(collect_set(user_ip ) ) as IP , 0 ,0 , 0 ,0 from logtable
where day='20170526' and hour='19';
```

计算结果如图 12-17 所示。

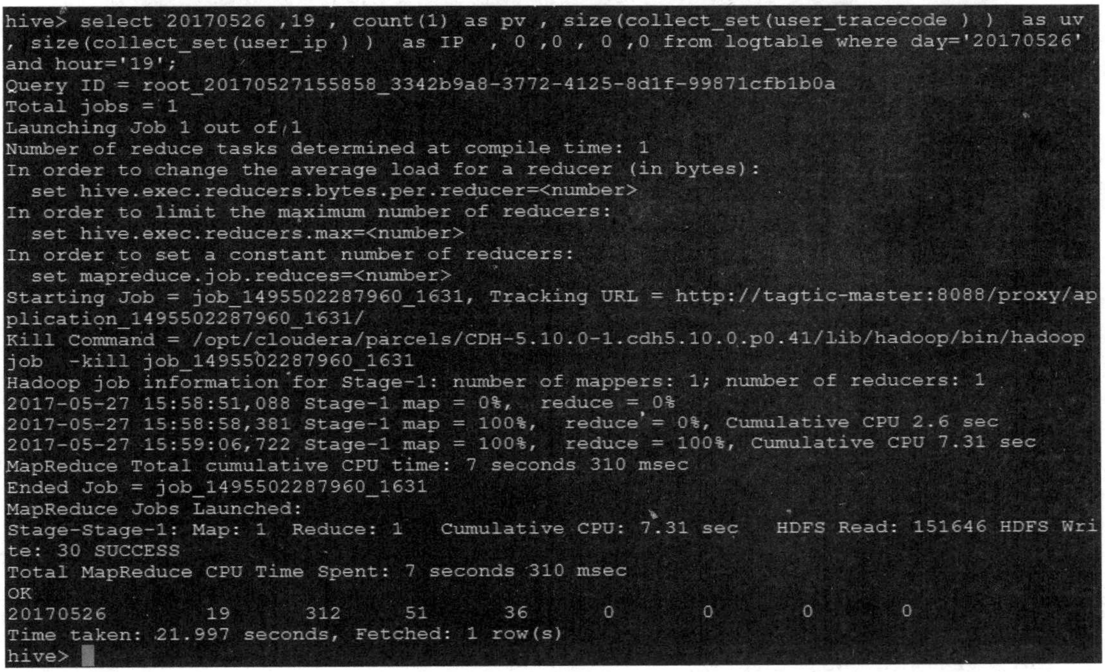

```
hive> select 20170526 ,19 , count(1) as pv , size(collect_set(user_tracecode ) ) as uv
, size(collect_set(user_ip ) ) as IP , 0 ,0 , 0 ,0 from logtable where day='20170526'
and hour='19';
Query ID = root_20170527155858_3342b9a8-3772-4125-8d1f-99871cfb1b0a
Total jobs = 1
Launching Job 1 out of 1
Number of reduce tasks determined at compile time: 1
In order to change the average load for a reducer (in bytes):
  set hive.exec.reducers.bytes.per.reducer=<number>
In order to limit the maximum number of reducers:
  set hive.exec.reducers.max=<number>
In order to set a constant number of reducers:
  set mapreduce.job.reduces=<number>
Starting Job = job_1495502287960_1631, Tracking URL = http://tagtic-master:8088/proxy/ap
plication_1495502287960_1631/
Kill Command = /opt/cloudera/parcels/CDH-5.10.0-1.cdh5.10.0.p0.41/lib/hadoop/bin/hadoop
job  -kill job_1495502287960_1631
Hadoop job information for Stage-1: number of mappers: 1; number of reducers: 1
2017-05-27 15:58:51,088 Stage-1 map = 0%,  reduce = 0%
2017-05-27 15:58:58,381 Stage-1 map = 100%,  reduce = 0%, Cumulative CPU 2.6 sec
2017-05-27 15:59:06,722 Stage-1 map = 100%,  reduce = 100%, Cumulative CPU 7.31 sec
MapReduce Total cumulative CPU time: 7 seconds 310 msec
Ended Job = job_1495502287960_1631
MapReduce Jobs Launched:
Stage-Stage-1: Map: 1  Reduce: 1   Cumulative CPU: 7.31 sec   HDFS Read: 151646 HDFS Wri
te: 30 SUCCESS
Total MapReduce CPU Time Spent: 7 seconds 310 msec
OK
20170526        19      312     51      36      0       0       0       0
Time taken: 21.997 seconds, Fetched: 1 row(s)
hive>
```

图 12-17　计算结果

Eclipse 查询 Hive 结果如图 12-18 所示。

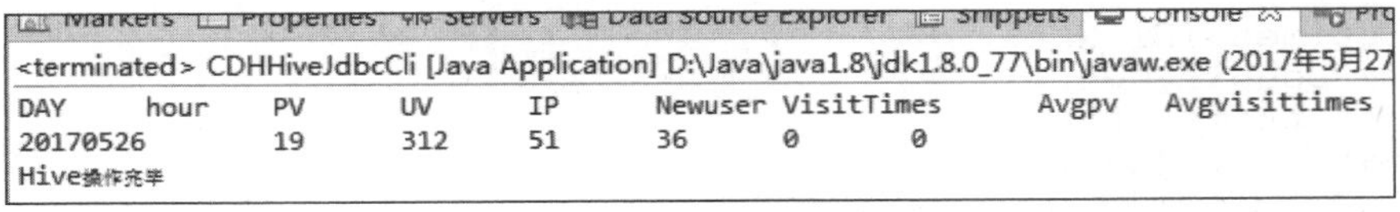

```
Markers  Properties  Servers  Data Source Explorer  Snippets  Console  Pro
<terminated> CDHHiveJdbcCli [Java Application] D:\Java\java1.8\jdk1.8.0_77\bin\javaw.exe (2017年5月27
DAY       hour   PV    UV    IP    Newuser VisitTimes    Avgpv   Avgvisittimes
20170526         19    312   51    36      0       0
Hive操作完毕
```

图 12-18　计算结果

7. 结果插入到 MySQL 中

连接 MySQL 界面如图 12-19 所示，查看 MySQL 表格结果如图 12-20 所示。

```
[root@tagtic-master ~]# mysql -uroot -ptagtic-master
Warning: Using a password on the command line interface can be insecure.
Welcome to the MySQL monitor.  Commands end with ; or \g.
Your MySQL connection id is 3405
Server version: 5.6.32 MySQL Community Server (GPL)

Copyright (c) 2000, 2016, Oracle and/or its affiliates. All rights reserved.

Oracle is a registered trademark of Oracle Corporation and/or its
affiliates. Other names may be trademarks of their respective
owners.

Type 'help;' or '\h' for help. Type '\c' to clear the current input statement

mysql>
```

图 12-19　连接 MySQL

```
mysql> show databases;
+--------------------+
| Database           |
+--------------------+
| information_schema |
| amon               |
| cm                 |
| druid              |
| druidtest          |
| hive               |
| hue                |
| mysql              |
| oozie              |
| performance_schema |
| yuhui              |
+--------------------+
```

图 12-20　查看 MySQL 表格结果

建立 MySQL 表格：

```
CREATE TABLE logtable(
    id int not null auto_increment,
    day VARCHAR(45) null,
    hour VARCHAR(45) null,
    pv INT NULL,
    uv INT NULL,
    ip INT NULL,
    newuser INT NULL,
    visittimes INT NULL,
    avgpv Double NULL,
    avgvisittimes Double NULL,
    primary key (id ,day ,hour)
)ENGINE=InnoDB DEFAULT CHARSET=utf8;
```

结果如图 12-21 所示。

```
mysql> use yuhui;
Database changed
mysql> show tables;
+-----------------+
| Tables_in_yuhui |
+-----------------+
| logtable        |
+-----------------+
1 row in set (0.00 sec)
```

图 12-21　查询表

程序运行完成之后，MySQL 查看结果，如图 12-22 所示。

id	day	hour	pv	uv	ip	newuser	visittimes	avgpv	avgvisittimes
1	2017052	19	312	51	36	0	0	0	0
2	2017052	20	165	21	22	0	0	0	0

图 12-22　查看数据

8. 前台页面展示

前台页面展示如图 12-23 所示。

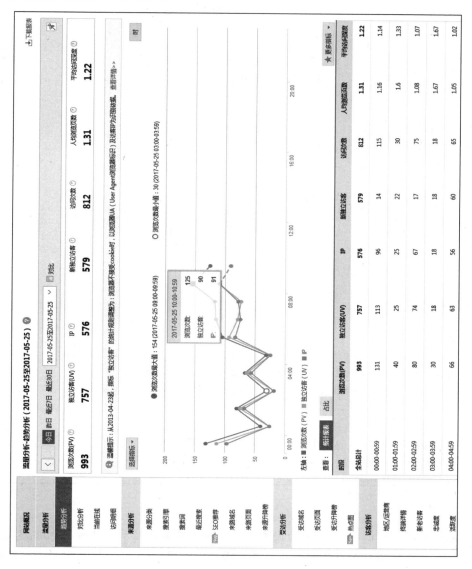

图 12-23　前台页面展示

9. 自动化任务

（1）Shell 定时任务

每小时的第 8 分钟进行自动计算。

```
[root@tongji ~]# crontab -l
30 * * * * ntpdate time.nist.gov
8 */1 * * * /home/yuhui/tendencyanalyze.sh
```

（2）Shell 获取当前时间的前一个小时

自动清洗日志，自动追加日志，程序日志保存到本地，便于查看。

```
[root@tongji yuhui ]# cat tendencyanalyze.sh
#!/bin/sh

SHELL=/bin/sh
PATH=/sbin:/bin:/usr/sbin:/usr/bin
MAILTO=root

#上一个小时的时间
s1=`date +%Y%m%d%H -d '-1 hours'`

#建立查看日志
touch /home/yuhui/tendency_hour_$s1.log

logfilepath=/home/yuhui/tendency_hour_$s1.log
echo $s1
echo $s1 >> $logfilepath
day=${s1:0:8}
hour=${s1:8:10}

#日志清洗
echo clean log $s1 begining............... >> $logfilepath
hadoop jar /home/yuhui/LogClean.jar $s1
echo clean log $s1 end............... >> $logfilepath

#hive 添加数据到分区
hive -e "ALTER TABLE logtable ADD IF NOT EXISTS
PARTITION(day='${day}',hour='${hour}') LOCATION
'/user/yuhui/cleanlogs/${day}/${hour}';"
```

10. 业务扩展

- 地域分析：可以将 IP 转换为国家、省、市，在页面显示用户所在地址。
- 鼠标轨迹：用户会有鼠标单击事件，而事件是规则的 X 轴和 Y 轴展示。提取事件，通过 HTML5 前端技术显示鼠标走势。

- 热力图：通过 IP 获得的国家、省、市，可以集体定位那些城市。用户登录本网站，在地图上面，颜色较深的则为用户多者。

12.3 网站实时项目

12.3.1 业务框架图

实时项目业务框架如图 12-24 所示。

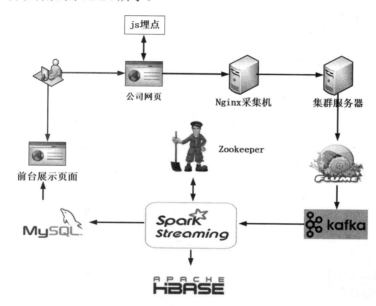

图 12-24　实时项目业务框架

12.3.2 子服务"当前在线"详解

1. 页面展示

当前在线页面如图 12-25、图 12-26 所示。

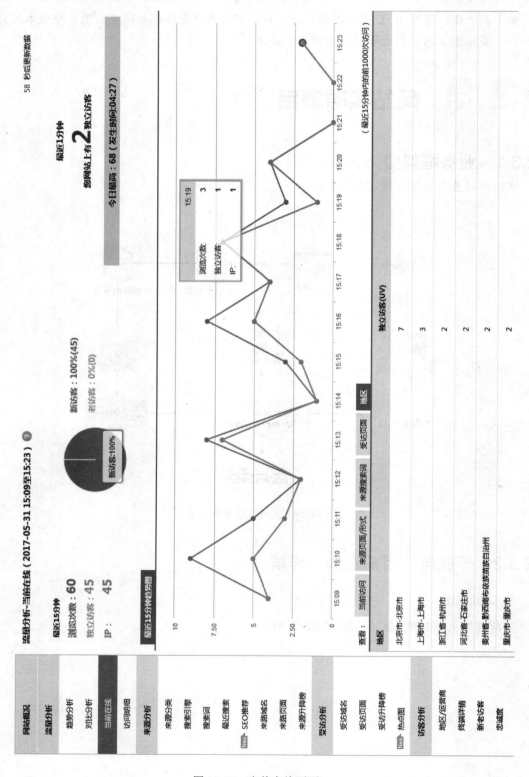

图 12-25 当前在线页面一

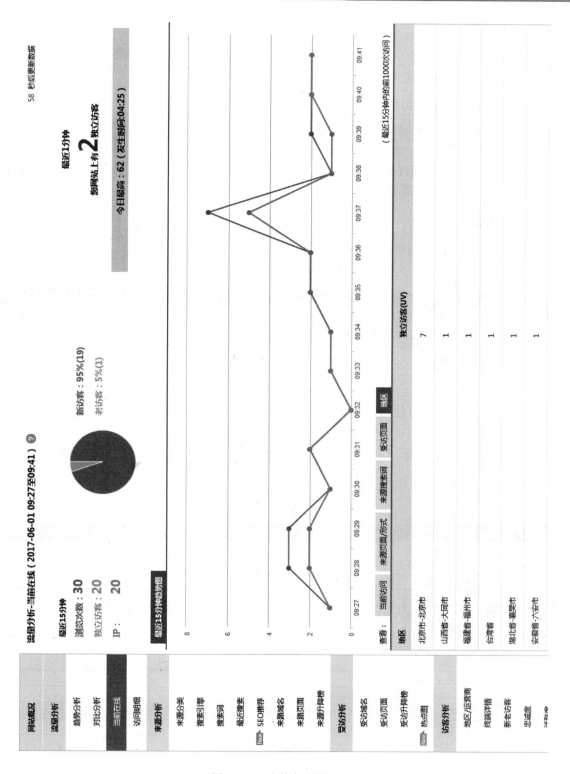

图 12-26　当前在线页面二

2. 功能说明

根据最近 15 分钟，提供网站用户数量，通过用户数量的趋势变化形态，可帮助你分析出网站访客的访问规律和用户在线状况等。

（1）了解网站质量和运营状况

如果网站用户整体偏低，说明网站内容不足以吸引访客，或者网站运营不够好。

（2）及时发现异常，避免流量继续下跌

例如在子服务"当前在线"中发现当前用户数量异常下跌，于是可以立即排查用户数量下跌原因，避免网站利益继续受损。

3. 字段说明

字段说明如表 12-5 所示。

表 12-5 字段说明

字段	说明
浏览次数(PV)	即通常说的 PV（PageView）值，用户每打开 1 个网站页面，记录 1 个 PV。用户多次打开同一页面 PV 累计多次
独立访客(UV)	1 天（00:00~24:00）之内，访问网站的不重复用户数（以浏览器 Cookie 为依据），一天内同一访客多次访问网站只被计算 1 次
IP	1 天（00:00~24:00）之内，访问网站的不重复 IP 数。一天内相同 IP 地址多次访问网站只被计算 1 次
地区	最近 15 分钟之内，访客的地址

4. 采集日志字段说明

采集日志字段说明如表 12-6 所示。

表 12-6 采集日志字段说明

字段	描述
suuid	suuid 信息
timestamp	时间戳 2016-02-01T03:02:14.222Z
request_method	请求类型（get、post、put、delete）
status	访问状态码（数字） eg：200
referer	上次访问的 URL
user_agent	浏览器 user agent
IP	用户的 IP 地址
http_url	访问当前 URL
to_target	下一个目标 URL
staytime	上次访问 URL 的停留时间
event	事件：duration（页面关闭发送，也能不发送）、jump（页面跳转）、load（加载页面）
short_cookie	过期时间为 10 分钟，每次都不相同
is_new	1:是 0：不是
appkey	topic name
page_id	每个页面的唯一 id

5. 转换日志说明

在业务处理中，我们会从采集日志将数据进行清洗和转换，表 12-7 所示的是根据采集日志转换而成的，字段来源在取值一列中有讲解，具体转换过程本章中不做过多讲解。

表 12-7 转换日志字段说明

行业 HDFS 字段 最终字段	类型	web 取值	取值范围
event	string	都是小写字符	注册（register） 加载（startup） 跳转（jump） 页面停留（shutdown） 事件标签类型（event） 页面标签类型（page） 消费（consumption） 会话持续时间（visit_duration）
event_name	Array	用户自定义	
account	string	NULL	
zone_id	string	NULL	
use_duration_cat	string	use_duration 的枚举值	0~10 秒、11~30 秒、31 秒~1 分钟、1~3 分钟、3~10 分钟、10~30 分钟、30~60 分钟、60 分钟以上
use_interval_cat	string	use_interval 的枚举值	首次、1 天内、1~2 天、2~3 天、3~5 天、5~7 天、7 天以上
use_duration	int	页面访问时长	
use_interval	string	一次回话的时长	
app_version	string	NULL	
device_type	string	NULL	
network	string	NULL	
nettype	string	NULL	
lang	string	NULL	
display	string	NULL	
register_days	int	NULL	
os_upgrade_from	string	NULL	
app_upgrade_from	string	NULL	
suuid	string	MD5(IP+UA)	
short_cookie	string	sdk 生成，当在有效时间范围内有新的访问时，需要更新失效时间	
url	string	http_url	
is_entrance	string	0 或 1,0 为非入口页	
data_type	string	Web 填 jump 类型相关的值	

（续表）

行业 HDFS 字段		web	行业 HDFS 字段
to_target	int	统计浏览量，通过 refer 获取	
search_engine	string	通过 http_referer 中的信息确定	
keyword	string	通过 http_referer 中的信息确定	
is_exit	int	标识是否是退出页	0 或者 1
domain	string	通过参数传递	
to_domain	int	通过 to_target 获得	
browser_version	string	通过 user agent 获得	
browser	string	通过 user agent 获得	
ip	string	通过 ip 获得	
day	string	通过 timestamp 转换获得	格式：2017-01-01
pv_cat	string	会话内 pv 次数的枚举值	1、2、3、4、5~10、10 以上
refer	string	通过 refer 获得	
timestamp	string	通过 timestamp 获得	格式：2017-01-01T01:24:18.336Z
appkey	string	通过 appkey 获得	
channel	string	NULL	
country	string	通过 ip 转换获得	
area	string	通过 ip 转换获得	
province	string	通过 ip 转换获得	
city	string	通过 ip 转换获得	
hour	string	通过 timestamp 转换获得	
gender	string	NULL	
age	int	NULL	
account_level	int	NULL	
payment_method	string	NULL	
consumption_point	string	NULL	
money	double	NULL	
os	string	通过 user agent 获得	
os_type	string	通过 user agent 获得	

12.3.3 表格的设计

1. HBase 表格字段设计

HBase 表格字段描述，如表 12-8 所示。

表 12-8　HBase 表格字段说明

字段名称	字段名称	字段类型
timestamp	时间蹉	Varchar
Hour	小时	Varchar
Minute	分钟	Varchar
Second	秒	Varchar
suuid	Suuid	Varchar
ip	ip	Varchar
Country	国家	Varchar
area	地区	Varchar
province	省	Varchar
city	市	Varchar

2. HBase 表格建立

```
hbase(main)> create 'RealTime',{NAME => 'RealTimeLogs', VERSIONS => 2}
```

3. MySQL 表格字段设计

MySQL 表格字段描述，如表 12-9 所示。

表 12-9　MySQL 表格字段说明

字段	字段名称	字段类型
day	天	String
hour	小时	String
minute	分钟	String
second	秒	String
suuid	Suuid	String
ip	ip	String
country	国家	String
area	地区	String
province	省	String
city	市	String

4. MySQL 表格建立

```
CREATE TABLE realTime (
    id int not null auto_increment,
    day VARCHAR(45) null,
    hour VARCHAR(45) null,
    suuid VARCHAR(45) null,
    minute VARCHAR(45) null,
    second VARCHAR(45) null,
    ip VARCHAR(45) null,
    country VARCHAR(45) null,
    area VARCHAR(45) null,
    province VARCHAR(45) null,
    city VARCHAR(45) null,
    primary key (id ,day ,hour, suuid)
```

```
)ENGINE=InnoDB DEFAULT CHARSET=utf8;
```

12.3.4 提前准备

网站实时项目的编辑器是 IntillJ IDEA，编程语言是 Scala，环境是 CDH5.10.0。在指标计算中只计算 id、day、hour、suuid、minute、second、ip、country、area、province、city，其余指标计算读者可以自己操作，本机主要以数据走向和框架介绍为主。

1. 下载客户端配置

参考 12.2.4 小节的内容配置好客户端。

2. 项目代码展示

项目代码展示如图 12-27 所示。

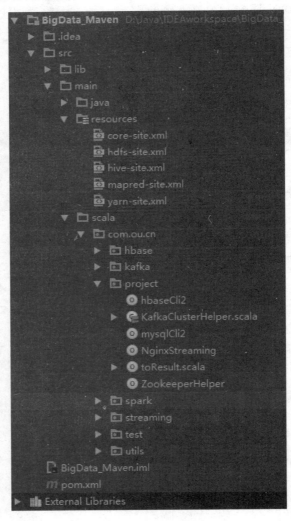

图 12-27　项目代码展示

3. IntelliJ IDEA 的 POM 文件

```xml
<project xmlns="http://maven.apache.org/POM/4.0.0"
xmlns:xsi="http://www.w3.org/2001/XMLSchema-instance"
xsi:schemaLocation="http://maven.apache.org/POM/4.0.0
http://maven.apache.org/maven-v4_0_0.xsd">
  <modelVersion>4.0.0</modelVersion>
  <groupId>BigData_Maven</groupId>
  <artifactId>BigData_Maven</artifactId>
  <version>1.0-SNAPSHOT</version>
  <inceptionYear>2008</inceptionYear>
  <properties>
    <scala.version>2.7.0</scala.version>
  </properties>

  <repositories>
    <repository>
      <id>scala-tools.org</id>
      <name>Scala-Tools Maven2 Repository</name>
      <url>http://scala-tools.org/repo-releases</url>
    </repository>
  </repositories>

  <pluginRepositories>
    <pluginRepository>
      <id>scala-tools.org</id>
      <name>Scala-Tools Maven2 Repository</name>
      <url>http://scala-tools.org/repo-releases</url>
    </pluginRepository>
  </pluginRepositories>

  <dependencies>

    <!-- https://mvnrepository.com/artifact/org.apache.spark/spark-sql_2.11 -->
    <dependency>
      <groupId>org.apache.spark</groupId>
      <artifactId>spark-sql_2.11</artifactId>
      <version>1.6.1</version>
      <!--<scope>provided</scope>-->
    </dependency>
    <!-- https://mvnrepository.com/artifact/org.apache.spark/spark-streaming_2.11 -->
    <dependency>
      <groupId>org.apache.spark</groupId>
      <artifactId>spark-streaming_2.11</artifactId>
      <version>1.6.1</version>
      <!--<scope>provided</scope>-->
    </dependency>
    <!--
https://mvnrepository.com/artifact/org.apache.phoenix/phoenix-spark -
```

```xml
-->
    <dependency>
      <groupId>org.apache.phoenix</groupId>
      <artifactId>phoenix-spark</artifactId>
      <version>4.7.0-HBase-1.1</version>
    </dependency>
    <dependency>
      <groupId>org.apache.spark</groupId>
      <artifactId>spark-streaming-kafka_2.11</artifactId>
      <version>1.6.1</version>
    </dependency>

    <dependency>
      <groupId>com.alibaba</groupId>
      <artifactId>fastjson</artifactId>
      <version>1.2.11</version>
    </dependency>
    <dependency>
      <groupId>org.apache.httpcomponents</groupId>
      <artifactId>httpclient</artifactId>
      <version>4.3.3</version>
    </dependency>
    <dependency>
      <groupId>commons-cli</groupId>
      <artifactId>commons-cli</artifactId>
      <version>1.2</version>
    </dependency>
    <!-- https://mvnrepository.com/artifact/org.jdbi/jdbi -->
    <dependency>
      <groupId>org.jdbi</groupId>
      <artifactId>jdbi</artifactId>
      <version>2.75</version>
    </dependency>
    <!-- https://mvnrepository.com/artifact/com.mchange/c3p0 -->
    <dependency>
      <groupId>com.mchange</groupId>
      <artifactId>c3p0</artifactId>
      <version>0.9.5.2</version>
    </dependency>

    <!-- https://mvnrepository.com/artifact/com.fasterxml.jackson.module/jackson-module-scala_2.11 -->
    <dependency>
      <groupId>com.fasterxml.jackson.module</groupId>
      <artifactId>jackson-module-scala_2.11</artifactId>
      <version>2.4.2</version>
    </dependency>
    <!-- https://mvnrepository.com/artifact/it.sauronsoftware.cron4j/cron4j --
```

```xml
>
  <dependency>
    <groupId>it.sauronsoftware.cron4j</groupId>
    <artifactId>cron4j</artifactId>
    <version>2.2.5</version>
  </dependency>
  <dependency>
    <groupId>io.vertx</groupId>
    <artifactId>vertx-web</artifactId>
    <version>3.2.1</version>
  </dependency>
  <dependency>
    <groupId>org.apache.kafka</groupId>
    <artifactId>kafka-clients</artifactId>
    <version>0.9.0.1</version>
  </dependency>
  <dependency>
    <groupId>org.apache.kafka</groupId>
    <artifactId>kafka_2.11</artifactId>
    <version>0.8.2.1</version>
  </dependency>
  <dependency>
    <groupId>com.typesafe.play</groupId>
    <artifactId>play-json_2.11</artifactId>
    <version>2.5.4</version>
  </dependency>
  <dependency>
    <groupId>org.scalaj</groupId>
    <artifactId>scalaj-http_2.11</artifactId>
    <version>2.3.0</version>
  </dependency>
  <dependency>
    <groupId>org.apache.spark</groupId>
    <artifactId>spark-hive_2.11</artifactId>
    <version>1.6.1</version>
  </dependency>
  <dependency>
    <groupId>redis.clients</groupId>
    <artifactId>jedis</artifactId>
    <version>2.9.0</version>
  </dependency>
</dependencies>

<build>
  <sourceDirectory>src/main/scala</sourceDirectory>
  <testSourceDirectory>src/test/scala</testSourceDirectory>
  <plugins>
    <plugin>
      <groupId>org.scala-tools</groupId>
      <artifactId>maven-scala-plugin</artifactId>
```

```xml
      <executions>
        <execution>
          <goals>
            <goal>compile</goal>
            <goal>testCompile</goal>
          </goals>
        </execution>
      </executions>
      <configuration>
        <scalaVersion>${scala.version}</scalaVersion>
        <args>
          <arg>-target:jvm-1.5</arg>
        </args>
      </configuration>
    </plugin>
    <plugin>
      <groupId>org.apache.maven.plugins</groupId>
      <artifactId>maven-eclipse-plugin</artifactId>
      <configuration>
        <downloadSources>true</downloadSources>
        <buildcommands>
          <buildcommand>ch.epfl.lamp.sdt.core.scalabuilder</buildcommand>
        </buildcommands>
        <additionalProjectnatures>
          <projectnature>ch.epfl.lamp.sdt.core.scalanature</projectnature>
        </additionalProjectnatures>
        <classpathContainers>
          <classpathContainer>org.eclipse.jdt.launching.JRE_CONTAINER</classpathContainer>
          <classpathContainer>ch.epfl.lamp.sdt.launching.SCALA_CONTAINER</classpathContainer>
        </classpathContainers>
      </configuration>
    </plugin>
  </plugins>
</build>
<reporting>
  <plugins>
    <plugin>
      <groupId>org.scala-tools</groupId>
      <artifactId>maven-scala-plugin</artifactId>
      <configuration>
        <scalaVersion>${scala.version}</scalaVersion>
      </configuration>
    </plugin>
  </plugins>
```

```
    </reporting>
</project>
```

4. 建立 scala 连接 HBase 的类

```scala
package com.ou.cn.project

import org.apache.hadoop.conf.Configuration
import org.apache.hadoop.hbase.client._
import org.apache.hadoop.hbase.{HBaseConfiguration,
HColumnDescriptor, HTableDescriptor, TableName}

/**
  * Created by yuhui on 2017/6/1 0001.
  */
object hbaseCli2 {

  val conf : Configuration= HBaseConfiguration.create
  conf.set("hbase.zookeeper.property.clientPort", "2181")
  conf.set("hbase.zookeeper.quorum", "tagtic-master,tagtic-slave02,tagtic-slave03")

  //创建表
  def createHTable(connection: Connection,tablename: String): Unit=
  {
    //Hbase 表模式管理器
    val admin = connection.getAdmin
    //本例将操作的表名
    val tableName = TableName.valueOf(tablename)
    //如果需要创建表
    if (!admin.tableExists(tableName)) {
      //创建 Hbase 表模式
      val tableDescriptor = new HTableDescriptor(tableName)
      //创建列簇1    RealTimeLogs
      tableDescriptor.addFamily(new
HColumnDescriptor("RealTimeLogs".getBytes()))
      //创建表
      admin.createTable(tableDescriptor)
      println("create done.")
    }
  }

  def insetLogs(logs : ResultLog ): Unit ={
    try{
      //Connection 的创建是个重量级的工作，线程安全，是操作 hbase 的入口
      val connection= ConnectionFactory.createConnection(conf)
      //创建表测试
      try {
        val Rowkey = logs.timestamp+'\t'+logs.suuid+'\t'
insertHTable(connection,"RealTime","RealTimeLogs","timestamp",Rowkey,
```

```
logs.timestamp)
insertHTable(connection,"RealTime","RealTimeLogs","day",Rowkey,logs.day)
insertHTable(connection,"RealTime","RealTimeLogs","hour",Rowkey,logs.hour)
insertHTable(connection,"RealTime","RealTimeLogs","minute",Rowkey,logs.minute)
insertHTable(connection,"RealTime","RealTimeLogs","second",Rowkey,logs.second)
insertHTable(connection,"RealTime","RealTimeLogs","suuid",Rowkey,logs.suuid)
insertHTable(connection,"RealTime","RealTimeLogs","ip",Rowkey,logs.ip)
insertHTable(connection,"RealTime","RealTimeLogs","country",Rowkey,logs.country)
insertHTable(connection,"RealTime","RealTimeLogs","area",Rowkey,logs.area)
insertHTable(connection,"RealTime","RealTimeLogs","province",Rowkey,logs.province)
insertHTable(connection,"RealTime","RealTimeLogs","city",Rowkey,logs.city)
      }finally {
        connection.close

      }
    }
  }

  //插入记录
  def insertHTable(connection:Connection,tablename:String,family:String,column:String,key:String,value:String):Unit={
    try{
      val userTable = TableName.valueOf(tablename)
      val table=connection.getTable(userTable)
      //准备 key 的数据
      val p=new Put(key.getBytes)
      //为 put 操作指定 column 和 value
      p.addColumn(family.getBytes,column.getBytes,value.getBytes())
      //提交一行
```

```
      table.put(p)
    }
  }

  def main(args: Array[String]): Unit = {
    // val sparkConf = new SparkConf().setAppName("HBaseTest")
    //启用spark上下文,只有这样才能驱动spark并行计算框架
    //val sc = new SparkContext(sparkConf)
    //创建一个配置,采用的是工厂方法
    val conf = HBaseConfiguration.create
    conf.set("hbase.zookeeper.property.clientPort", "2181")
    conf.set("hbase.zookeeper.quorum", "tagtic-master,tagtic-slave02,tagtic-slave03")

    try{
      //Connection 的创建是个重量级的工作,线程安全,是操作hbase的入口
      val connection= ConnectionFactory.createConnection(conf)
      //创建表测试
      try {
        createHTable(connection, "RealTime")
        //插入数据,重复执行为覆盖

insertHTable(connection,"RealTime","RealTimeLogs","Hadoop","002","Hadoop for me")

insertHTable(connection,"RealTime","RealTimeLogs","Hadoop","003","Java for me")

insertHTable(connection,"RealTime","RealTimeLogs","Spark","002","Scala for me")

      }finally {
        connection.close
        //   sc.stop
      }
    }
  }
}
```

5. 建立 KafkaClusterHelper 类

```
package com.ou.cn.project

/**
  * Created by yuhui on 16-6-29.
  * copy from spark-kafka source
  */
```

```scala
import java.util.Properties

import kafka.api._
import kafka.common.{ErrorMapping, TopicAndPartition}
import kafka.consumer.{ConsumerConfig, SimpleConsumer}
import org.apache.spark.SparkException

import scala.collection.mutable.ArrayBuffer
import scala.util.Random
import scala.util.control.NonFatal


/**
 * Convenience methods for interacting with a Kafka cluster.
 *
 * @param kafkaParams Kafka <a
 href="http://kafka.apache.org/documentation.html#configuration">
 *                    configuration parameters</a>.
 *                    Requires "metadata.broker.list" or
 "bootstrap.servers" to be set with Kafka broker(s),
 *                    NOT zookeeper servers, specified in
 host1:port1,host2:port2 form
 */
class KafkaClusterHelper(val kafkaParams: Map[String, String]) extends Serializable {

  import KafkaClusterHelper.{Err, LeaderOffset, SimpleConsumerConfig}

  // ConsumerConfig isn't serializable
  @transient private var _config: SimpleConsumerConfig = null

  def config: SimpleConsumerConfig = this.synchronized {
    if (_config == null) {
      _config = SimpleConsumerConfig(kafkaParams)
    }
    _config
  }

  def connect(host: String, port: Int): SimpleConsumer =
    new SimpleConsumer(host, port, config.socketTimeoutMs,
      config.socketReceiveBufferBytes, config.clientId)

  def findLeaders(
              topicAndPartitions: Set[TopicAndPartition]
            ): Either[Err, Map[TopicAndPartition, (String, Int)]]
= {
    val topics = topicAndPartitions.map(_.topic)
    val response = getPartitionMetadata(topics).right
    val answer = response.flatMap { tms: Set[TopicMetadata] =>
```

```scala
      val leaderMap = tms.flatMap { tm: TopicMetadata =>
        tm.partitionsMetadata.flatMap { pm: PartitionMetadata =>
          val tp = TopicAndPartition(tm.topic, pm.partitionId)
          if (topicAndPartitions(tp)) {
            pm.leader.map { l =>
              tp -> (l.host -> l.port)
            }
          } else {
            None
          }
        }
      }.toMap

      if (leaderMap.keys.size == topicAndPartitions.size) {
        Right(leaderMap)
      } else {
        val missing = topicAndPartitions.diff(leaderMap.keySet)
        val err = new Err
        err.append(new SparkException(s"Couldn't find leaders for ${missing}"))
        Left(err)
      }
    }
    answer
  }

  def getPartitions(topics: Set[String]): Either[Err, Set[TopicAndPartition]] = {
    getPartitionMetadata(topics).right.map { r =>
      r.flatMap { tm: TopicMetadata =>
        tm.partitionsMetadata.map { pm: PartitionMetadata =>
          TopicAndPartition(tm.topic, pm.partitionId)
        }
      }
    }
  }

  def getPartitionMetadata(topics: Set[String]): Either[Err, Set[TopicMetadata]] = {
    val req = TopicMetadataRequest(
      TopicMetadataRequest.CurrentVersion, 0, config.clientId, topics.toSeq)
    val errs = new Err
    withBrokers(Random.shuffle(config.seedBrokers), errs) { consumer =>
      val resp: TopicMetadataResponse = consumer.send(req)
      val respErrs = resp.topicsMetadata.filter(m => m.errorCode != ErrorMapping.NoError)

      if (respErrs.isEmpty) {
```

```scala
          return Right(resp.topicsMetadata.toSet)
      } else {
        respErrs.foreach { m =>
          val cause = ErrorMapping.exceptionFor(m.errorCode)
          val msg = s"Error getting partition metadata for
'${m.topic}'. Does the topic exist?"
          errs.append(new SparkException(msg, cause))
        }
      }
    }
    Left(errs)
  }


  //获取 kafka 最新的 offset
  def getLatestLeaderOffsets(
                      topicAndPartitions: Set[TopicAndPartition]
                    ): Either[Err, Map[TopicAndPartition,
LeaderOffset]] =
    getLeaderOffsets(topicAndPartitions, OffsetRequest.LatestTime)

  def getEarliestLeaderOffsets(
                      topicAndPartitions: Set[TopicAndPartition]
                    ): Either[Err, Map[TopicAndPartition,
LeaderOffset]] =
    getLeaderOffsets(topicAndPartitions, OffsetRequest.EarliestTime)

  def getLeaderOffsets(
                   topicAndPartitions: Set[TopicAndPartition],
                   before: Long
                 ): Either[Err, Map[TopicAndPartition,
LeaderOffset]] = {
    getLeaderOffsets(topicAndPartitions, before, 1).right.map { r =>
      r.map { kv =>
        // mapValues isnt serializable, see SI-7005
        kv._1 -> kv._2.head
      }
    }
  }

  private def flip[K, V](m: Map[K, V]): Map[V, Seq[K]] =
    m.groupBy(_._2).map { kv =>
      kv._1 -> kv._2.keys.toSeq
    }

  def getLeaderOffsets(
                   topicAndPartitions: Set[TopicAndPartition],
                   before: Long,
                   maxNumOffsets: Int
                 ): Either[Err, Map[TopicAndPartition,
```

```scala
    Seq[LeaderOffset]]] = {
    findLeaders(topicAndPartitions).right.flatMap { tpToLeader =>
      val leaderToTp: Map[(String, Int), Seq[TopicAndPartition]] = flip(tpToLeader)
      val leaders = leaderToTp.keys
      var result = Map[TopicAndPartition, Seq[LeaderOffset]]()
      val errs = new Err
      withBrokers(leaders, errs) { consumer =>
        val partitionsToGetOffsets: Seq[TopicAndPartition] =
          leaderToTp((consumer.host, consumer.port))
        val reqMap = partitionsToGetOffsets.map { tp: TopicAndPartition =>
          tp -> PartitionOffsetRequestInfo(before, maxNumOffsets)
        }.toMap
        val req = OffsetRequest(reqMap)
        val resp = consumer.getOffsetsBefore(req)
        val respMap = resp.partitionErrorAndOffsets
        partitionsToGetOffsets.foreach { tp: TopicAndPartition =>
          respMap.get(tp).foreach { por: PartitionOffsetsResponse =>
            if (por.error == ErrorMapping.NoError) {
              if (por.offsets.nonEmpty) {
                result += tp -> por.offsets.map { off =>
                  LeaderOffset(consumer.host, consumer.port, off)
                }
              } else {
                errs.append(new SparkException(
                  s"Empty offsets for ${tp}, is ${before} before log beginning?"))
              }
            } else {
              errs.append(ErrorMapping.exceptionFor(por.error))
            }
          }
        }
        if (result.keys.size == topicAndPartitions.size) {
          return Right(result)
        }
      }
      val missing = topicAndPartitions.diff(result.keySet)
      errs.append(new SparkException(s"Couldn't find leader offsets for ${missing}"))
      Left(errs)
    }
  }

  // Consumer offset api
  // scalastyle:off
  // https://cwiki.apache.org/confluence/display/KAFKA/A+Guide+To+The+Kafka+Protocol#AGuideToTheKafkaProtocol-OffsetCommit/FetchAPI
```

```scala
    // scalastyle:on

    // this 0 here indicates api version, in this case the original ZK backed api.
    private def defaultConsumerApiVersion: Short = 0

    // Try a call against potentially multiple brokers, accumulating errors
    private def withBrokers(brokers: Iterable[(String, Int)], errs: Err)
                           (fn: SimpleConsumer => Any): Unit = {
      brokers.foreach { hp =>
        var consumer: SimpleConsumer = null
        try {
          consumer = connect(hp._1, hp._2)
          fn(consumer)
        } catch {
          case NonFatal(e) =>
            errs.append(e)
        } finally {
          if (consumer != null) {
            consumer.close()
          }
        }
      }
    }

    //获取 kafka 最开始的 offset
    def getFromOffsets(kafkaParams: Map[String, String], topics: Set[String]): Map[TopicAndPartition, Long] = {
      val reset = kafkaParams.get("auto.offset.reset").map(_.toLowerCase)
      val result = for {
        topicPartitions <- getPartitions(topics).right
        leaderOffsets <- (if (reset == Some("smallest")) {
          getEarliestLeaderOffsets(topicPartitions)
        } else {
          getLatestLeaderOffsets(topicPartitions)
        }).right
      } yield {
        leaderOffsets.map { case (tp, lo) =>
          (tp, lo.offset)
        }
      }
      KafkaClusterHelper.checkErrors(result)
    }
}

object KafkaClusterHelper {
```

```scala
  type Err = ArrayBuffer[Throwable]

  /** If the result is right, return it, otherwise throw
SparkException */
  def checkErrors[T](result: Either[Err, T]): T = {
    result.fold(
      errs => throw new SparkException(errs.mkString("\n")),
      ok => ok
    )
  }


  case class LeaderOffset(host: String, port: Int, offset: Long)

  /**
    * High-level kafka consumers connect to ZK.  ConsumerConfig
assumes this use case.
    * Simple consumers connect directly to brokers, but need many of
the same configs.
    * This subclass won't warn about missing ZK params, or presence
of broker params.
    */

  class SimpleConsumerConfig private(brokers: String, originalProps:
Properties)
      extends ConsumerConfig(originalProps) {
    val seedBrokers: Array[(String, Int)] = brokers.split(",").map
{ hp =>
      val hpa = hp.split(":")
      if (hpa.size == 1) {
        throw new SparkException(s"Broker not in the correct format of
<host>:<port> [$brokers]")
      }
      (hpa(0), hpa(1).toInt)
    }
  }


  object SimpleConsumerConfig {
    /**
      * Make a consumer config without requiring group.id or
zookeeper.connect,
      * since communicating with brokers also needs common settings
such as timeout
      */
    def apply(kafkaParams: Map[String, String]): SimpleConsumerConfig
= {
      // These keys are from other pre-existing kafka configs for
specifying brokers, accept either
      val brokers = kafkaParams.get("metadata.broker.list")
```

```
      .orElse(kafkaParams.get("bootstrap.servers"))
      .getOrElse(throw new SparkException(
        "Must specify metadata.broker.list or bootstrap.servers"))

    val props = new Properties()
    kafkaParams.foreach { case (key, value) =>
      // prevent warnings on parameters ConsumerConfig doesn't know about
      if (key != "metadata.broker.list" && key != "bootstrap.servers") {
         props.put(key, value)
      }
    }

    Seq("zookeeper.connect", "group.id").foreach { s =>
      if (!props.containsKey(s)) {
        props.setProperty(s, "")
      }
    }

    new SimpleConsumerConfig(brokers, props)
  }
}
```

6. 建立 Scala 连接 MySQL 的类

```
package com.ou.cn.project

import java.sql.{Connection, DriverManager}

/**
  * Created by Administrator on 2017/6/1 0001.
  */
object mysqlCli2 {

  // 驱动程序名
  val driver : String = "com.mysql.jdbc.Driver"

  // URL 指向要访问的数据库名 scutcs
  val url : String = "jdbc:mysql://tagtic-master:3306/yuhui?useUnicode=true&characterEncoding=utf8";

  // MySQL 配置时的用户名
  val user : String = "root"

  // MySQL 配置时的密码
  val password : String= "tagtic-master"
```

```scala
  def main(args: Array[String]): Unit = {

    var log = "{\"timestamp\":\"2017-02-11T10:45:04.004Z\",\"url\":\"/idonews/\",\"is_entrance\":0,\"data_type\":\"null\",\"channel\":\"null\",\"to_target\":0,\"keyword\":\"\",\"search_engine\":\"\",\"country\":\"中国\",\"area\":\"华东\",\"province\":\"浙江省\",\"city\":\"杭州市\",\"use_duration_cat\":\"null\",\"domain\":\"www.donews.com\",\"to_domain\":0,\"use_interval_cat\":\"null\",\"is_exit\":0,\"event\":\"startup\",\"os\":\"Windows XP\",\"os_type\":\"pc\",\"browser\":\"Internet Explorer\",\"browser_version\":\"Internet Explorer 8.0\",\"suuid\":\"81f3d97da1aba500f928ea677892d689\",\"short_cookie\":\"null\",\"ip\":\"122.225.222.98\",\"use_duration\":0,\"use_interval\":0,\"pv_cat\":\"null\",\"event_name\":[],\"refer\":\"\",\"hour\":\"10\",\"gender\":\"null\",\"age\":0,\"account_level\":0,\"payment_method\":\"\",\"consumption_point\":\"\",\"money\":0.0,\"account\":\"\",\"zone_id\":\"\",\"app_version\":\"\",\"network\":\"null\",\"nettype\":\"null\",\"lang\":\"\",\"app_upgrade_from\":\"\",\"display\":\"null\",\"device_type\":\"null\",\"register_days\":0,\"refer_domain\":\"null\",\"appkey\":\"donews_website_nginx_log\",\"day\":\"2017-02-11\"}"

    InsertSql(toResult.toResultObjec(log))

  }

  def InsertSql(resultLog: ResultLog): Unit ={

    try{

      var conn:Connection = null

      Class.forName(driver);

      conn = DriverManager.getConnection(url, user, password);

      val stmt = conn.createStatement();

      val insql="insert into realTime(day,hour,minute,second,suuid,ip,country,area,province,city) values(?,?,?,?,?,?,?,?,?,?)";

      val ps=conn.prepareStatement(insql);

      ps.setString(1, resultLog.day);

      ps.setString(2, resultLog.hour);

      ps.setString(3, resultLog.minute);
```

```
        ps.setString(4, resultLog.second);

        ps.setString(5, resultLog.suuid);

        ps.setString(6, resultLog.ip);

        ps.setString(7, resultLog.country);

        ps.setString(8, resultLog.area);

        ps.setString(9, resultLog.province);

        ps.setString(10, resultLog.city);

        ps.executeUpdate();

        System.out.println("mysql 数据插入完毕");

        stmt.close();

        conn.close();

      } catch {
        case e => e.printStackTrace
      }

    }

}
```

7. 建立 NginxStreaming 类

```
package cn.orcale.com.project;

import java.sql.*;

public class mysqlCli2 {

    // 驱动程序名
    static String driver = "com.mysql.jdbc.Driver";

    // URL 指向要访问的数据库名 scutcs
    static String url = "jdbc:mysql://tagtic-master:3306/yuhui";

    // MySQL 配置时的用户名
    static String user = "root";

    // MySQL 配置时的密码
    static String password = "tagtic-master";
```

```java
//      public static void main(String[] args) throws Exception{
//
//          mysqlCli mysqlCli = new mysqlCli();
//          Logs logs = new Logs();
//          logs.DAY="20170527";
//          logs.HOUR="14";
//          logs.PV=100;
//          logs.UV=200;
//          logs.IP=300;
//          logs.Newuser=400;
//          logs.VisitTimes=500;
//          logs.Avgpv=500.0;
//          logs.Avgvisittimes=600.0;
//          mysqlCli2.InsertSql(logs);
//
//      }

    public static void InsertSql(Logs log){

        try {

            Connection conn = null ;

                Class.forName(driver);

                conn = DriverManager.getConnection(url, user, password);

                Statement stmt = conn.createStatement();

                String insql="insert into logtable(day,hour,pv,uv,ip,newuser,visittimes,Avgpv,Avgvisittimes) "
                    + "values(?,?,?,?,?,?,?,?,?)";

                PreparedStatement ps=conn.prepareStatement(insql);

                ps.setString(1, log.getDAY());

                ps.setString(2, log.getHOUR());

                ps.setInt(3, log.getPV());

                ps.setInt(4, log.getUV());

                ps.setInt(5, log.getIP());

                ps.setInt(6, log.getNewuser());

                ps.setInt(7, log.getVisitTimes());

                ps.setDouble(8, log.getAvgpv());
```

```
                ps.setDouble(9, log.getAvgvisittimes());

                ps.executeUpdate();

                System.out.println("mysql 数据插入完毕");

                stmt.close();
                conn.close();


        } catch (Exception e) {
            e.printStackTrace();
        }

    }

}
```

8. 建立 Log 对象的 case 类

```
package com.ou.cn.project

import com.fasterxml.jackson.databind.ObjectMapper
import com.fasterxml.jackson.module.scala.DefaultScalaModule
import org.apache.hadoop.conf.Configuration
import org.apache.hadoop.hbase.{HBaseConfiguration, TableName}
import org.apache.hadoop.hbase.client.{Connection, ConnectionFactory, Put}

/**
 * Created by Administrator on 2017/6/1 0001.
 */

case class ResultLog(
                var timestamp : String,
                var day : String,
                var hour : String,
                var minute : String,
                var second : String,
                var suuid : String,
                var ip : String,
                var country : String,
                var area : String,
                var province : String,
                var city: String
            )

case class NginxLog(
                var timestamp: String,
                var url : String,
```

```scala
    var is_entrance : java.lang.Integer,
    var data_type: String,
    var channel: String,

    var to_target: java.lang.Integer,
    var keyword: String,
    var search_engine: String,
    var country : String,
    var area : String,

    var province : String,
    var city : String,
    var use_duration_cat : String,
    var domain : String,
    var to_domain : java.lang.Integer,

    var use_interval_cat : String,
    var is_exit : java.lang.Integer,
    var event : String,
    var os : String,
    var os_type : String,


    var browser : String,
    var browser_version : String,
    var suuid : String,
    var short_cookie : String,
    var ip : String,

    var use_duration : java.lang.Integer,
    var use_interval : java.lang.Integer,
    var pv_cat : String,
    var event_name : Array[String],
    var refer : String,

    var hour : java.lang.Integer,
    var gender : String,
    var age : java.lang.Integer,
    var account_level : java.lang.Integer,
    var payment_method : String,

    var consumption_point : String,
    var money: Double,
    var account : String,
    var zone_id : String,
    var app_version : String,

    var network : String,
    var nettype : String,
    var lang : String,
```

```scala
                    var  app_upgrade_from : String,
                    var  display : String,

                    var  device_type : String,
                    var  register_days: java.lang.Integer,
                    var  refer_domain : String,
                    var  appkey : String,
                    var  day : String

                    )


object toResult {

  def main(args: Array[String]): Unit = {

    var log = "{\"timestamp\":\"2017-02-11T11:22:42.042Z\",\"url\":\"/%7b%7bnew.url%7d%7d\",\"is_entrance\":0," +

"\"data_type\":\"null\",\"channel\":\"null\",\"to_target\":0,\"keyword\":\"\",\"search_engine\":\"\"," +
      "\"country\":\"美国\",\"area\":\"0\",\"province\":\"0\",\"city\":\"0\",\"use_duration_cat\":\"null\"," +

"\"domain\":\"www.donews.com\",\"to_domain\":0,\"use_interval_cat\":\"null\",\"is_exit\":0," +

"\"event\":\"startup\",\"os\":\"Linux\",\"os_type\":\"pc\",\"browser\":\"Firefox\"," +
      "\"browser_version\":\"Firefox 3.0.3\",\"suuid\":\"30f0dbc17223fc2037ee0ea6ca1e9772\"," +

"\"short_cookie\":\"null\",\"ip\":\"69.30.236.186\",\"use_duration\":0,\"use_interval\":0," +

"\"pv_cat\":\"null\",\"event_name\":[],\"refer\":\"\",\"hour\":\"11\",\"gender\":\"null\"," +

"\"age\":0,\"account_level\":0,\"payment_method\":\"\",\"consumption_point\":\"\",\"money\":0.0," +

"\"account\":\"\",\"zone_id\":\"\",\"app_version\":\"\",\"network\":\"null\",\"nettype\":\"null\"," +

"\"lang\":\"\",\"app_upgrade_from\":\"\",\"display\":\"null\",\"device_type\":\"null\"," +

"\"register_days\":0,\"refer_domain\":\"null\",\"appkey\":\"donews_we
```

```
bsite_nginx_log\",\"day\":\"2017-02-11\"}";
    val logs : ResultLog = toResultObjec(log)


  }
  //插入记录
  def insertHTable(connection:Connection,tablename:String,family:String,column:String,key:String,value:String):Unit={
    try{
      val userTable = TableName.valueOf(tablename)
      val table=connection.getTable(userTable)
      //准备 key 的数据
      val p=new Put(key.getBytes)
      //为 put 操作指定 column 和 value
      p.addColumn(family.getBytes,column.getBytes,value.getBytes())
      //提交一行
      table.put(p)
    }
  }


  def toResultObjec(log :String): ResultLog ={

    val mapper = new ObjectMapper()

    mapper.registerModule(DefaultScalaModule)

    val obj = mapper.readValue(log, classOf[NginxLog])

    val  rs = ResultLog("","","","","","","","","","","")

    rs.timestamp = obj.timestamp
    rs.day = obj.day
    rs.hour = obj.timestamp.substring(11,13)
    rs.minute = obj.timestamp.substring(14,16)
    rs.second = obj.timestamp.substring(17,19)
    rs.suuid = obj.suuid
    rs.ip = obj.ip
    rs.country = obj.country
    rs.area = obj.area
    rs.province = obj.province
    rs.city= obj.city
    rs

  }

}
```

9. 建立 Scala 连接 ZooKeeper 类

```scala
package com.ou.cn.project
import kafka.common.TopicAndPartition
import org.apache.curator.framework.CuratorFrameworkFactory
import org.apache.curator.retry.ExponentialBackoffRetry
import org.slf4j.LoggerFactory

import scala.collection.JavaConversions._

/**
 * Created by yuhui on 16-6-8.
 */
object ZookeeperHelper {

  val ZOOKEEPER_CONNECT="tagtic-master:2181,tagtic-slave02:2181,tagtic-slave03:2181"

  val LOG = LoggerFactory.getLogger(ZookeeperHelper.getClass)
  val client = {
    val client = CuratorFrameworkFactory
      .builder
      .connectString(ZOOKEEPER_CONNECT)
      .retryPolicy(new ExponentialBackoffRetry(1000, 3))
      .namespace("ou")
      .build()
    client.start()
    client
  }

  //zookeeper 创建路径
  def ensurePathExists(path: String): Unit = {
    if (client.checkExists().forPath(path) == null) {
      client.create().creatingParentsIfNeeded().forPath(path)
    }
  }

  //zookeeper 加载 offset 的方法
  def loadOffsets(topicSet: Set[String], defaultOffset: Map[TopicAndPartition, Long]): Map[TopicAndPartition, Long] = {
    val kafkaOffsetPath = s"/kafkaOffsets"
    ensurePathExists(kafkaOffsetPath)
    val offsets = for {
    //t 就是路径 webstatistic/kafkaOffsets 下面的子目录遍历
      t <- client.getChildren.forPath(kafkaOffsetPath)
      if topicSet.contains(t)
    //p 就是新路径   /webstatistic/kafkaOffsets/donews_website
      p <- client.getChildren.forPath(s"$kafkaOffsetPath/$t")
    } yield {
      //遍历路径下面的 partition 中的 offset
      val data = client.getData.forPath(s"$kafkaOffsetPath/$t/$p")
      //将 data 变成 Long 类型
      val offset = java.lang.Long.valueOf(new String(data)).toLong
      (TopicAndPartition(t, Integer.parseInt(p)), offset)
    }
    defaultOffset ++ offsets.toMap
  }

  //zookeeper 存储 offset 的方法
```

```
def storeOffsets(offsets: Map[TopicAndPartition, Long]): Unit = {
  val kafkaOffsetPath = s"/kafkaOffsets"
  if (client.checkExists().forPath(kafkaOffsetPath) == null) {
client.create().creatingParentsIfNeeded().forPath(kafkaOffsetPath)
  }
  for ((tp, offset) <- offsets) {
    val data = String.valueOf(offset).getBytes
    val path = s"$kafkaOffsetPath/${tp.topic}/${tp.partition}"
    ensurePathExists(path)
    client.setData().forPath(path, data)
  }
}
```

10. 使用组件及软件

Flume、Kafka、SparkStreaming、HBase、Mysql、Intellj Idea、Scala、ZooKeeper。

12.3.5 项目步骤

1. 日志埋点

本步骤请参考 12.2.5 小节离线项目中的日志埋点步骤。

2. 数据采集服务器

Nginx 服务器收集原始数据，每 5 分钟切割文件，并且发送到 Flume 日志采集服务器。同时每天零点将昨天的日志进行压缩保存，日志名一定要规范按照 YYYY-MM-DD.log，便于以后容错机制的日志恢复。

3. 字段转换

实时项目中使用的日志是从采集日志中进行转换过来的，本节中不讲转换过程，只用转换好的结果日志进行项目的实战操作。转换好的日志字段请参考【转换日志说明】。

4. 日志展示

项目中不讲日志转换过程，项目过程中只通过转换好的结果数据，进行业务的处理。
处理好的数据展示：

```
{"timestamp":"2017-02-11T10:42:42.042Z","url":"/net/201501/2877461.shtm","is_entrance":0,"data_type":"null","channel":"null","to_target":0,"keyword":"","search_engine":"","country":"中国","area":"华东","province":"山东省","city":"青岛市","use_duration_cat":"null","domain":"www.donews.com","to_domain":0,"use_interval_cat":"null","is_exit":0,"event":"startup","os":"Windows 8","os_type":"pc","browser":"Chrome","browser_version":"Chrome 43.0.2357.65","suuid":"d0f79068944531b60c9647865e6e4752","short_cookie":"null","ip":"115.28.36.228","use_duration":0,"use_interval":0,"pv_cat":"null","event_name":[],"refer":"","hour":"10","gender":"null","a
```

```
ge":0,"account_level":0,"payment_method":"","consumption_point":"","m
oney":0.0,"account":"","zone_id":"","app_version":"","network":"null"
,"nettype":"null","lang":"","app_upgrade_from":"","display":"null","d
evice_type":"null","register_days":0,"refer_domain":"null","appkey":"
donews website nginx log","day":"2017-02-11"}
```

处理好的结果数据如图 12-28 所示。

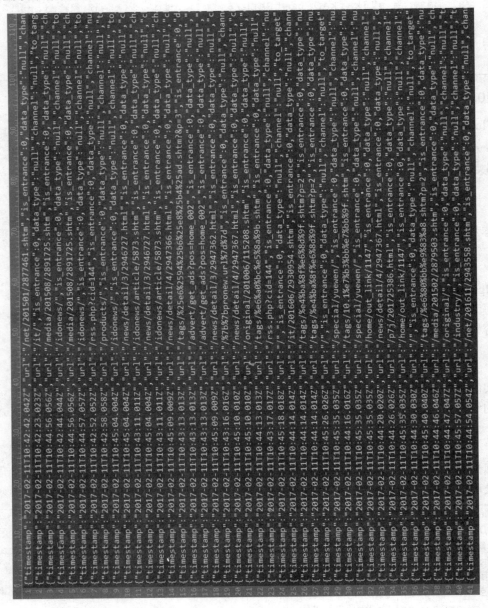

图 12-28 处理好的结果数据

5. Flume 配置

（1）Kafka 的 topic 查看

```
kafka-topics --list --zookeeper localhost:2181
```

（2）Kafka 的 topic 建立

```
kafka-topics --create --topic topicTest2 --replication-factor 2 --partitions 10 --zookeeper localhost:2181
```

（3）Flume 的路径建立

```
[root@tagtic-slave03 yuhui]# mkdir /home/yuhui/flumePath
[root@tagtic-slave03 yuhui]# mkdir /home/yuhui/flumePath/dir
[root@tagtic-slave03 yuhui]# mkdir /home/yuhui/flumePath/dir/logdfs
[root@tagtic-slave03 yuhui]# mkdir /home/yuhui/flumePath/dir/logdfstmp
[root@tagtic-slave03 yuhui]# mkdir /home/yuhui/flumePath/dir /point
[root@tagtic-slave03 yuhui]# chmod 777 -R /home/yuhui/flumePath/dir
```

（4）Flume 的配置文件

```
#agent1 name
agent1.sources=source1
agent1.sinks=sink1
agent1.channels=channel1


#Spooling Directory
#set source1
agent1.sources.source1.type=spooldir
agent1.sources.source1.spoolDir=/home/yuhui/flumePath/dir/logdfs
agent1.sources.source1.channels=channel1
agent1.sources.source1.fileHeader = false
agent1.sources.source1.interceptors = i1
agent1.sources.source1.interceptors.i1.type = timestamp

#set sink1
agent1.sinks.sink1.type = org.apache.flume.sink.kafka.KafkaSink
agent1.sinks.sink1.topic = topicTest2
agent1.sinks.sink1.brokerList = tagtic-slave01:9092,tagtic-slave02:9092,tagtic-slave03:9092
agent1.sinks.sink1.requiredAcks = 1
agent1.sinks.sink1.batchSize = 100
agent1.sinks.sink1.channel = channel1

#set channel1
agent1.channels.channel1.type=file
agent1.channels.channel1.checkpointDir=/home/yuhui/flumePath/dir/logdfstmp/point
agent1.channels.channel1.dataDirs=/home/yuhui/flumePath/dir/logdfstmp
```

（5）Kafka 的消费命令及且执行结果

```
kafka-console-consumer --zookeeper localhost:2181 --topic topicTest -
-from-beginning
```

Kafka 消费结果如图 12-29 所示。

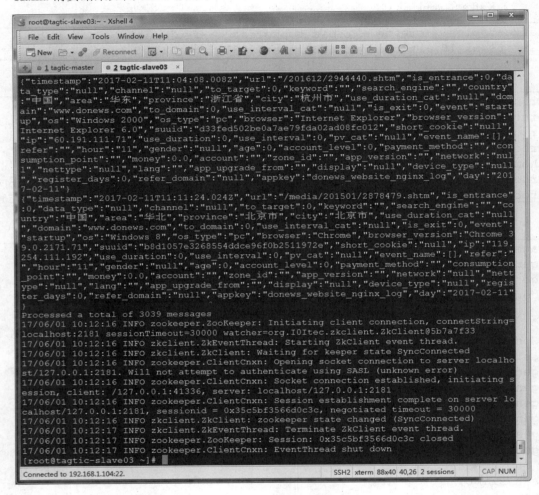

图 12-29　Kafka 消费结果

6. SparkStreaming 读写 Kafka

通过 NginxStreaming.scala 类读取 Kafka 的数据，且将结果分别发到 MySQL 和 HBase 中，执行主要代码如下：

```
//将数据分别插入到 HBase 和 MySQL 中
rdd.foreach(logs=>{
  mysqlCli2.InsertSql(toResult.toResultObjec(logs))
  hbaseCli2.insetLogs(toResult.toResultObjec(logs))
})
```

7. 无丢失获取 Kafka 数据

按照 NginxStreaming.scala 类进行讲解，无丢失获取 Kafka 数据是实时日志的核心，最为重要的一个环节。

需求说明如下：

- 将 Kafka 中的数据无丢失提取。
- 详解 Kafka 读取数据以及提取 Offset 的步骤。
- 详解 ZooKeeper 存储 TopicAndPartition 和对应的 Offset。

前期准备：

- 将 org.apache.Spark.streaming.kafka.KafkaCluster 类抽出来变成 KafkaClusterHelper。
- 建立 Scala 连接 ZooKeeper 类 ZookeeperHelper，便于存储 Topic 的 offset。

代码实现步骤说明如下。

（1）将 org.apache.spark.streaming.kafka.KafkaCluster 类抽出来变成 KafkaClusterHelper。

（2）编写 ZookeeperHelper 类便于将 TopicAndPartition 和对应的 Offset 存储到 ZooKeeper 中。

（3）使用 set("spark.streaming.kafka.maxRatePerPartition","1000")控制每次读取的 Kafka 数量，避免数据读取过大导致内存不足或者操作过慢。

（4）通过 ZookeeperHelper.loadOffsets 方法获取 Kafka 中 Topic 对应的 offset，如果第一次则为空，Kafka 从 0 开始读取；如果是第二次则获取 Topic 对应的 offset，从上次截止的 Offset 读起。

（5）通过 KafkaUtils.createDirectStream 方法获取 Topic 指定的 Offset 变成 DStream。

（6）将 DStream 中的 RDD 分别插入 Mysql 和 HBase 中。

（7）数据处理完毕，则将 Topic 对应的 offset 存到 ZooKeeper 中，ZooKeeper 存储 Offset。

（8）SparkStreaming 第二次则可以在 ZooKeeper 中获取上次的 Offset。

（9）循环第 4 步到第 8 步，圆圈是详细步骤，具体操作如图 12-30 所示。

无丢失获取 Kafka 数据步骤如图 12-30 所示。

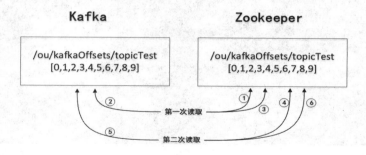

图 12-30　无丢失获取 Kafka 数据步骤

ZooKeeper 存储 Offset 路径结构如图 12-31 所示。

图 12-31　ZooKeeper 存储 Offse 路径结构

8. 结果数据进入 HBase

```
hbase(main):005:0> scan 'RealTime',{LIMIT=>1}
```

HBase 客户端查询结果如图 12-32 所示。

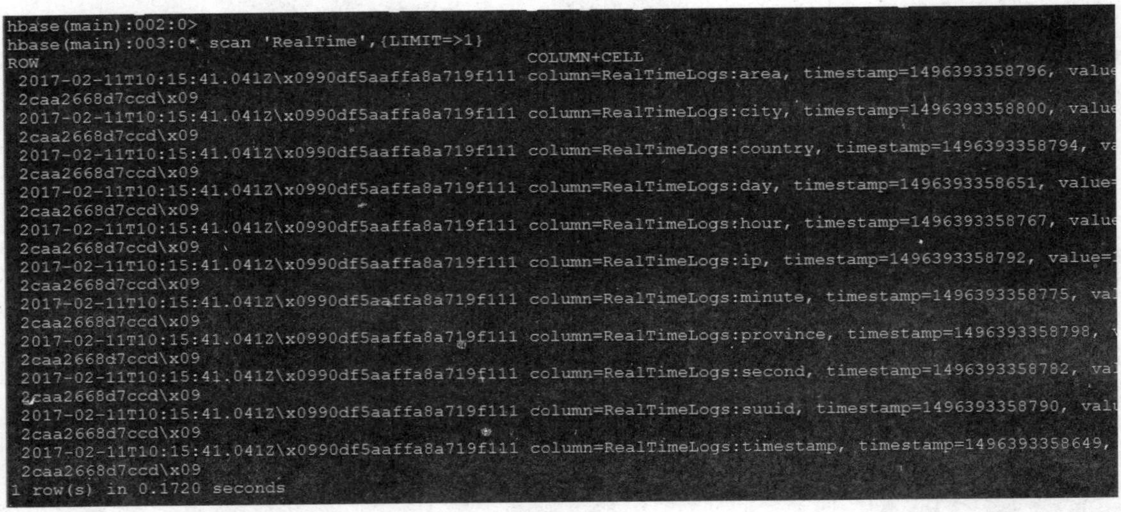

图 12-32　HBase 客户端查询结果

9. 结果数据进入 MySQL

（1）查看 MySQL 表格（如图 12-33、图 12-34 所示）

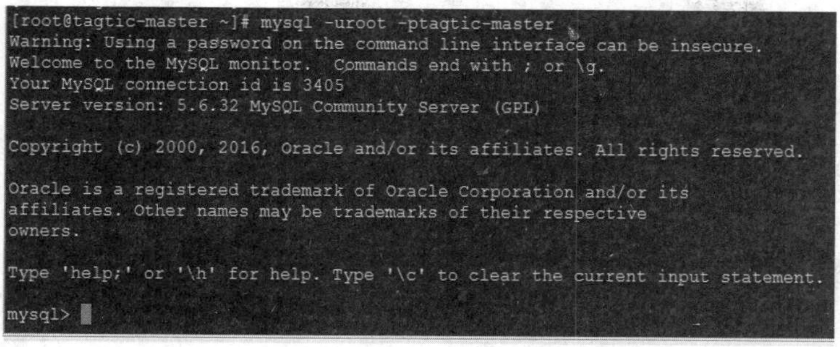

```
[root@tagtic-master ~]# mysql -uroot -ptagtic-master
Warning: Using a password on the command line interface can be insecure.
Welcome to the MySQL monitor.  Commands end with ; or \g.
Your MySQL connection id is 3405
Server version: 5.6.32 MySQL Community Server (GPL)

Copyright (c) 2000, 2016, Oracle and/or its affiliates. All rights reserved.

Oracle is a registered trademark of Oracle Corporation and/or its
affiliates. Other names may be trademarks of their respective
owners.

Type 'help;' or '\h' for help. Type '\c' to clear the current input statement.

mysql>
```

图 12-33　查看 MySQL 表格一

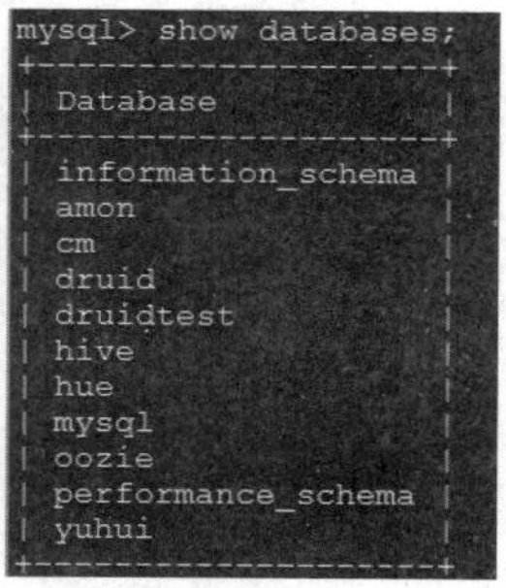

```
mysql> show databases;
+--------------------+
| Database           |
+--------------------+
| information_schema |
| amon               |
| cm                 |
| druid              |
| druidtest          |
| hive               |
| hue                |
| mysql              |
| oozie              |
| performance_schema |
| yuhui              |
+--------------------+
```

图 12-34　查看 MySQL 表格二

（2）建立 MySQL 表格

```
CREATE TABLE realTime (
    id int not null auto_increment,
    day VARCHAR(45) null,
    hour VARCHAR(45) null,
    suuid VARCHAR(45) null,
    minute VARCHAR(45) null,
    second VARCHAR(45) null,
    ip VARCHAR(45) null,
    country VARCHAR(45) null,
    area VARCHAR(45) null,
    province VARCHAR(45) null,
    city VARCHAR(45) null,
    primary key (id ,day ,hour, suuid)
 )ENGINE=InnoDB DEFAULT CHARSET=utf8;
```

结果如图 12-35 所示。

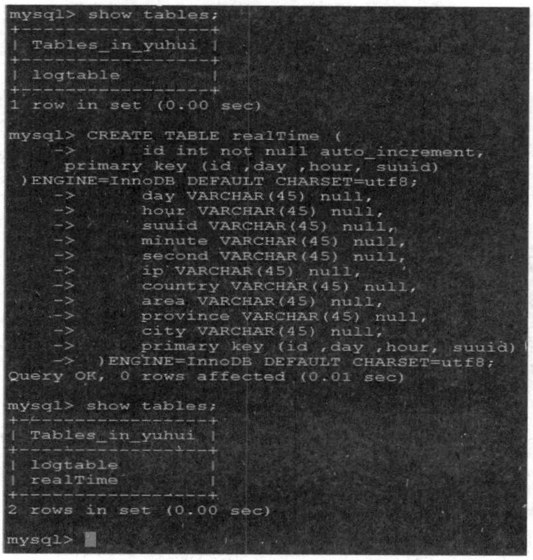

```
mysql> show tables;
+-----------------+
| Tables_in_yuhui |
+-----------------+
| logtable        |
+-----------------+
1 row in set (0.00 sec)

mysql> CREATE TABLE realTime (
    ->         id int not null auto_increment,
      primary key (id ,day ,hour, suuid)
  )ENGINE=InnoDB DEFAULT CHARSET=utf8;
    ->         day VARCHAR(45) null,
    ->         hour VARCHAR(45) null,
    ->         suuid VARCHAR(45) null,
    ->         minute VARCHAR(45) null,
    ->         second VARCHAR(45) null,
    ->         ip VARCHAR(45) null,
    ->         country VARCHAR(45) null,
    ->         area VARCHAR(45) null,
    ->         province VARCHAR(45) null,
    ->         city VARCHAR(45) null,
    ->         primary key (id ,day ,hour, suuid)
    ->    )ENGINE=InnoDB DEFAULT CHARSET=utf8;
Query OK, 0 rows affected (0.01 sec)

mysql> show tables;
+-----------------+
| Tables_in_yuhui |
+-----------------+
| logtable        |
| realTime        |
+-----------------+
2 rows in set (0.00 sec)

mysql>
```

图 12-35　建立 MySQL 表格

（3）程序运行完成之后，MySQL 查看结果（如图 12-36 所示）

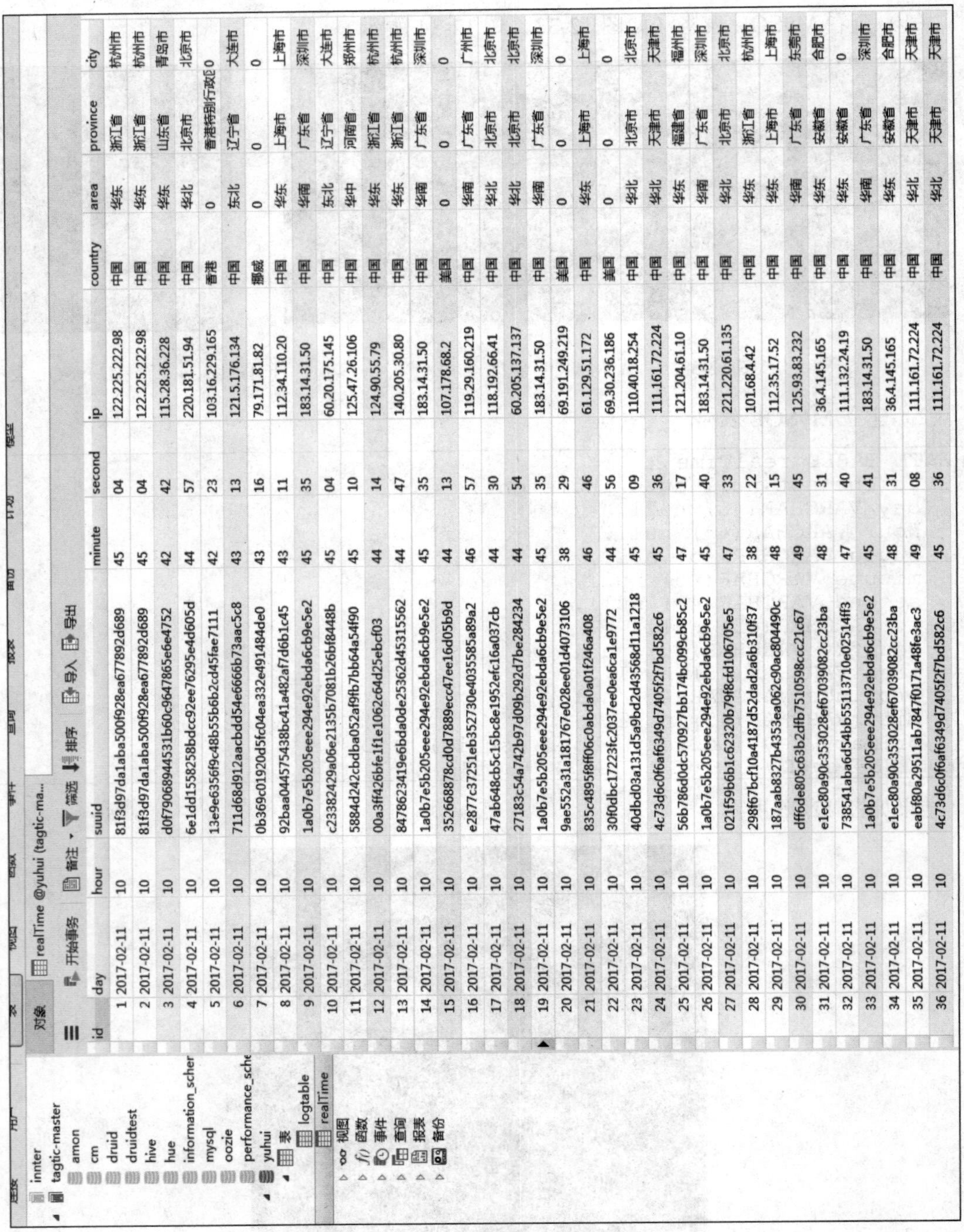

图 12-36　MySQL 查看结果

10. 前台展示

前台展示如图 12-37~图 12-38 所示。

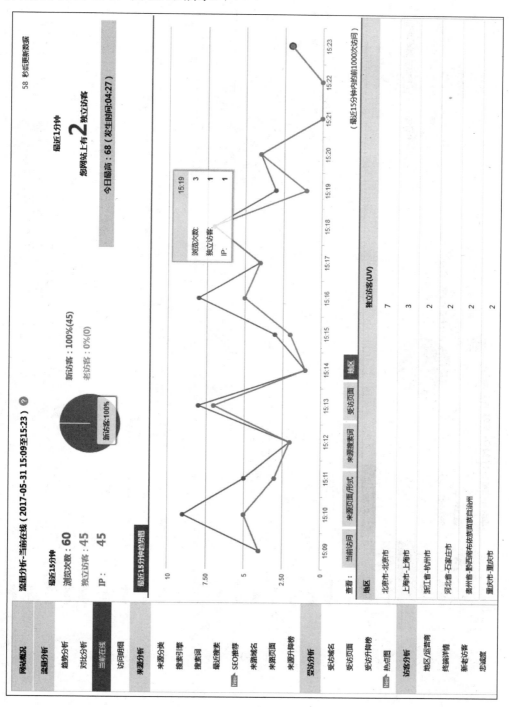

图 12-37

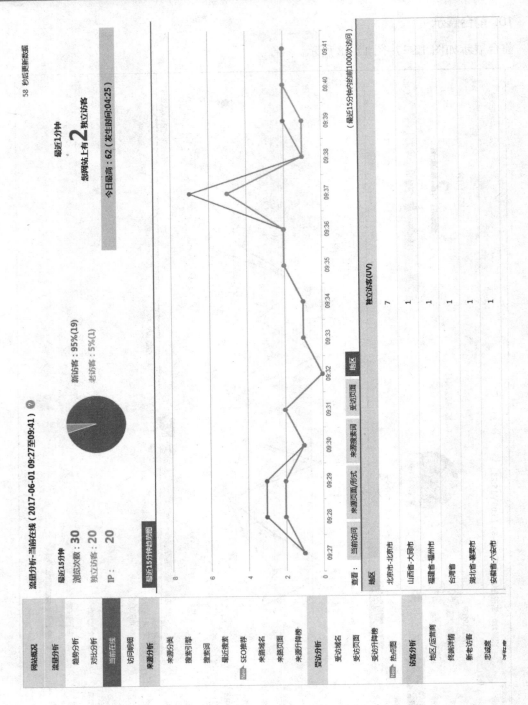

图 12-38

12.4 小结

本章重点讲解在 CDH 环境下市面上 2 个主流项目的实现，通过这两个项目的操作使读者更加清晰地掌握 Hadoop+Spark 的生态圈组件功能和操作。两个项目分别为离线项目和实时项目。

离线项目每小时进行数据采集、清洗、过滤、计算，最终结果推送到前台展现，使用到的组件为：HDFS、MapReduce、Hive、MySQLsql。

实时项目每 5 分钟进行数据采集、清洗、过滤、计算，最终结果推送到前台展现，使用到的组件为：Flume、Kafka、Spark Streaming、HBase、ZooKeeper、MySQLsql。